The Open-Ended Approach: A New Proposal for Teaching Mathematics

Edited by

Jerry P. Becker
Southern Illinois University
at Carbondale
United States

Shigeru Shimada
Science University
of Tokyo
Japan

NATIONAL COUNCIL OF TEACHERS OF MATHEMATICS
Reston, Virginia

Copyright © 1997 by
THE NATIONAL COUNCIL OF TEACHERS OF MATHEMATICS, INC.
1906 Association Drive, Reston, VA 20191-9988
www.nctm.org
Third printing 2000

Library of Congress Cataloging-in-Publication Data:

Sansū, sūgakka no ōpun endo apurōchi. English.
    The open-ended approach : a new proposal for teaching mathematics
/ edited by Jerry P. Becker, Shigeru Shimada.
        p.   cm.
    ISBN 0-87353-430-1 (paperback)
    1. Mathematics—Study and teaching—Japan.   I. Becker, Jerry P.
II. Shimada, Shigeru, 1916–   . III. Title.
QA14.J3S2513   1997
510'.71' 052—dc21   97-12708
                                                              CIP

The publications of the National Council of Teachers of Mathematics
present a variety of viewpoints. The views expressed or implied
in the publication, unless otherwise noted, should not be interpreted
as official positions of the Council.

Printed in the United States of America

Table of Contents

Preface to the Translation

WE ARE HAPPY to present this translation from Japanese to English of the book *Open-Ended Approach in Arithmetic and Mathematics: A New Proposal for the Improvement of Teaching*. The original book, edited by Shigeru Shimada, reported on Japanese developmental research concerned with methods of evaluating higher-order thinking in mathematics education using open-ended problems as a theme. The research led to an awareness that the "open ended" teaching approach used in the research had the potential to improve mathematics teaching and learning in Japanese schools.

The editors of this English translation of the book first met in the late 1960s. Their professional relationship grew over a number of years through scholarly interaction that involved correspondence, the exchange of materials, visits to Japan, and participation in international meetings there. At one of these meetings the idea of a United States–Japan seminar on mathematics education was discussed. This idea became a reality when delegations of mathematics educators from the two countries convened in 1986 at the East-West Center in Honolulu for the U.S.-Japan Seminar on Mathematical Problem Solving.

The seminar was productive and led to subsequent cross-national research that extended the interaction of mathematics educators in the two countries. This interaction led to a reciprocal and deepening understanding of the philosophy and practices of mathematics teaching and research in each country. The collaboration that unfolded after the seminar also helped considerably to remove the natural barriers of language and distance that existed. It paved the way for further substantial joint work that continues into the mid-1990s.

Japanese research, particularly that relating to the open-ended teaching approach, became one focus of the collaboration following the 1986 seminar. Subsequently, some of the contents of this book were discussed and an observation was made that its availability in English would be a useful addition to the scarce but slowly growing cross-cultural literature in mathematics education. Shigeru Shimada, Tatsuro Miwa, Toshio Sawada, and Yoshihiko Hashimoto supported the proposal to translate the original work. The editors and our colleague, Shigeo Yoshikawa, then undertook the work that produced this translation. Much of Yoshikawa's contribution was carried out while he was a visiting scholar at Southern Illinois University at Carbondale.

Research in problem solving in school mathematics was being addressed in Japan starting in the early 1970s. At the same time, problem solving was a focus of interest and research in the United States, represented, for example, by NCTM's *Agenda for Action* in 1980 and later by its *Curriculum and Evaluation Standards for School Mathematics* in 1989. Thus, the important role that problem solving plays in improving curriculum and instruction is widely recognized in both countries. Although this book clearly deals with important aspects of problem solving, we believe that it also has implications for improving assessment in problem solving, which is another important issue being addressed by researchers in both countries. Perhaps it will open a new approach to this fundamental issue that may influence beliefs, perspectives, and methodology in this major area of research and classroom practice. In this regard, the editors refer the reader to the paper "Classroom Assessment in Japanese Mathematics Education," by Eizo Nagasaki and Jerry P. Becker, which appears in NCTM's 1993 Yearbook, *Assessment in the Mathematics Classroom*.

This book reports on the first phase of Japanese research on the open-ended approach to teaching mathematics. The research was continued and further reported in the 1985 book *From Problem to Problem: The Improvement of Mathematics Teaching by the Developmental Treatment of Mathematical Problems,* edited by Yoshio Takeuchi and Toshio Sawada. This publication is also being translated into English as another contribution to the cross-cultural research literature in mathematics education.

This project would not have been possible without the significant support and efforts of several individuals. We wish to thank Tatsuro Miwa, Toshio Sawada, and Yoshihiko Hashimoto for their encouragement from the beginning. We also acknowledge and thank Shigeo Yoshikawa for his important and significant contribution in translating the first draft of this book, complemented by the significant work of the second editor. Our profound gratitude goes to Joan Griffin for her untiring and faithful efforts in typing the original and all edited drafts of the manuscript. We owe her our profound appreciation.

No external funds were used for translating the initial draft and preparing the final manuscript; even so, we wish to express appreciation to the National Science Foundation (NSF) and the Japan Society for the Promotion of Science (JSPS), which supported the 1986 seminar as well as the subsequent joint research from which this translation emerged. Finally, we express appreciation to the National Council of Teachers of Mathematics for publishing this book; to the NCTM Educational Materials Committee, the editorial staff, and the Board of Directors, we express gratitude both for their patience with us in completing the task and for making this book available to teachers, researchers, and the larger mathematics education community.

Preface to the Book

BETWEEN 1971 AND 1976, Japanese researchers carried out a series of developmental research projects on methods of evaluating higher-order-thinking skills in mathematics education. (The projects and the years during which they were conducted are as follows: 1971, Developmental Study of Methods of Evaluation in Mathematics Education; 1972–73, Developmental Study of Methods of Evaluation in Mathematics Education and Analysis of the Influence of Factors in Learning in Mathematics Education; 1974–76, Developmental Study of a Method of Evaluating Students' Achievement in Higher-Order-Thinking Skills in Mathematics Education.) The research was supported by a subsidy for scientific research from the Ministry of Education in Japan. This book is a partial outcome of the study.

In its first stage, the study's focus had been on substantiating the effectiveness of open-ended problems, as described in this book, as a method to evaluate higher-order-thinking skills. As the study progressed, however, we became aware that lessons based on solving open-ended problems as a central theme have a rich potential for improving teaching and learning, so we modified our efforts to address this conjecture also. Since we were able to confirm the conjecture through a series of practical teaching experiments, we decided to publish our conclusions in the form that follows. We summarized our findings as a proposal for the improvement of mathematics teaching, and we look to readers for constructive criticism.

The book is composed of seven chapters. Our basic philosophy is explained in chapter 1. The meaning of the unfamiliar phrase *open-ended approach* is illustrated by an example in chapter 2, and several points to be kept in mind while planning lessons are described in chapter 3. In chapters 4, 5, and 6, we give examples of open-ended problems that are used at various grade levels in the lessons developed and used by members of our study group. Throughout chapters 4 to 6, the format shown at the beginning of chapter 4 will be used to report the lesson carried out by the members. Chapter 7 is a summary of a round-table discussion by group members of their reflections on the study and on the follow-up of certain issues. Needless to say, we expect that the book will be read in the order of the chapters; however, starting at chapter 2 and proceeding to chapters 4, 5, or 6 and then going back to chapters 1 and 3 may be another useful approach.

All members of our study group participated in examining and discussing the outline and composition of each chapter. Each section or chapter was written by one of our members. I was responsible for the general editing of the entire work; Toshio Sawada, Yoshihiko Hashimoto, Sachino Kobayashi, and colleagues at the National Institute for Educational Research took care of preparing, editing, and proofreading the drafts.

The study was cooperatively carried out in close contact among subgroups in six areas: Yamagata, Chiba, Tokyo, Kanagawa, Fukui, and Oita. The names of members who contributed throughout the study are as follows:

Yoshio Takeuchi	Faculty of Education, Yamagata University
Masami Takasago	The Elementary School at Yamagata University
Hiroshi Ishiyama	The Lower Secondary School at Yamagata University
Shuei Sasaki	Yamagata Minami Upper Secondary School, Yamagata
Shiba Sugioka	Faculty of Education, Chiba University
Masakazu Aoyagi	Faculty of Education, Chiba University
Kanjiro Kobayashi	The Elementary School at Chiba University

Osamu Matsui	The Lower Secondary School at Chiba University
Yoshishige Sugiyama	Faculty of Education, Tokyo Gakugei University
Kiyoshi Takai	Matsudo Dental Faculty, Nihon University
Kozo Tsubota	Fukazawa Elementary School, Setagaya Ward, Tokyo
Yukio Yoshikawa	The Lower and Upper Secondary School at the University of Tokyo
Hiroshi Kimura	Faculty of Education, Yokohama National University
Eijyuro Matsubara	Ushioda Elementary School, Yokohama City
Tetsuko Kuwabara	Ushioda Elementary School, Yokohama City
Hiroaki Fukuju	Kamisugeta Lower Secondary School, Yokohama City
Zenko Ozawa	Yamakita Upper Secondary School, Kanagawa
Nobuhiko Nohda	Faculty of Education, Fukui University
Shizue Mori	Asahi Elementary School, Fukui City
Yasushi Aoyama	Shimin Lower Secondary School, Fukui City
Masashiro Hiroshima	The Lower Secondary School at Fukui University
Tetsuro Uemura	Faculty of Education, Oita University
Takehisa Asakura	Kamekawa Elementary School, Beppu City
Wataru Miyasako	Hamawaki Lower Secondary School, Beppu City
Shigeru Shimada	Science Education Research Center, National Institute for Educational Research
Toshio Sawada	Science Education Research Center, National Institute for Educational Research
Yoshihiko Hashimoto	Science Education Research Center, National Institute for Educational Research

Data were gathered in schools where members served and in other cooperating schools. Many other teachers in these schools cooperated with our members. Furthermore, still others who had an interest in our study provided data, suggestions, or encouragement. To all of them, we would like to express our deep gratitude for their cooperation and kindness.

Shigeru Shimada
August 1977

Chapter 1

The Significance of an Open-Ended Approach

SHIGERU SHIMADA

National Institute for Educational Research

TRADITIONAL PROBLEMS used in mathematics teaching in both elementary and secondary school classrooms have a common feature: that one and only one correct answer is predetermined. The problems are so well formulated that answers are either correct or incorrect (including incomplete ones) and the correct one is unique. We call these problems "complete" or "closed" problems.

WHAT IS AN OPEN-ENDED APPROACH?

We propose to call problems that are formulated to have multiple correct answers "incomplete" or "open ended" problems. Many examples of such problems can easily be found. In traditional classroom teaching, when students are asked to focus on and develop different methods, ways, or approaches to getting an answer to a given problem and not on finding the answer to the problem, the students are, in a sense, facing and dealing with an open-ended problem, since what is asked for is not the answer to the problem but rather the methods for arriving at an answer. Thus, there is not just one approach but several or many. In such instances, however, the "openness" is lost if the teacher proceeds as though only one method is presupposed as the correct one.

In the teaching method that we call an "open-ended approach," an "incomplete" problem is presented first. The lesson then proceeds by using many correct answers to the given problem to provide experience in finding something new in the process. This can be done through combining students' own knowledge, skills, or ways of thinking that have previously been learned.

Details of this new approach will be explained in later chapters. For some concrete examples, see chapter 2.

The purpose of this chapter is to explain our motivation for proposing a new and different approach and how this approach fits into the whole spectrum of the mathematics education of students.

MOTIVE AND PROCESS OF OUR STUDY

The problem we originally had in mind was how to evaluate students' achievement of the objectives of higher-order thinking (hereafter referred to as "higher objectives") in mathematics education. In mathematics teaching, a series of knowledge, skills, concepts, principles, or laws is presented to students in step-by-step fashion. This series is taught not because each item is considered important in isolation from the others but because we expect that the series will be integrated with the abilities and attitudes of each student, thereby forming an intellectual organization within the mind of each student. Although individual knowledge, skills, and such are important components of a whole, the essential point is that they should be integrated into the intellectual makeup of each student. Therefore, in order to know the extent to which students achieve higher objectives, we must observe how they use what is learned in a concrete situation as well as how they cope when what has been learned does not work directly. Such observation is easier said than done because the concrete situation needs to be in a natural context (rather than an artificial one created by others for the purpose of evaluation). Though such a situation may sometimes emerge naturally in the classroom or in students' daily lives, it would usually occur only incidentally.

In contrast, most paper-and-pencil tests that are used for data gathering in evaluation employ the closed type of problems. In such problems, all mathematical conditions needed for solution are completely furnished, and it is sufficient for students to retrieve their learned knowledge and skills and apply the appropriate ones to find the solutions by using the given problem conditions as a guide. Therefore, the evaluation cannot go beyond checking students' achievement in terms of their knowledge, skills, or capabilities of identifying and applying concepts, principles, or laws.

If achieving these measurable objectives is a sufficient condition for achieving the higher objectives, then an evaluation using closed problems could be used for assessing the higher objectives. Can it be assumed that the former is a sufficient condition for the latter, or at least that they are related to each other in such a way that the latter can be predicted with a high probability from the former? In an attempt to answer this question, we formed a group to study the following two questions:

1. What examples of students' behavior can be considered measures of the higher objectives?

Though it is difficult to evaluate directly the achievement of higher objectives during classroom teaching, what student behaviors can be regarded as being measures of them? In other words, what are the desirable behavior patterns that students show? Providing such examples is one of our tasks.

2. How are the observed desirable student behaviors that we consider measures of higher objectives related to students' achievement as measured by paper-and-pencil tests or other instruments?

In other words, can we expect that those students who perform well on ordinary tests also exhibit desirable behaviors and that improvement measured by tests is also accompanied by an improvement of observed behaviors?

For the first question, since we thought that no objective criteria for observing such behavior patterns existed, we sent a questionnaire to mathematics education leaders throughout Japan to gather opinions on this issue. Most concrete comments gathered could be summarized as follows: "In facing a problem situation, the students can mathematize the situation and deal with it," or in other words, "In analyzing a problem situation, the students bring forth an (important) aspect of the problem into their favored way of thinking by mobilizing their repertoire of learned mathematics, reinterpreting it to deal with the situation mathematically, and then applying their preferred technique."

In order to gather data for the second question, it was necessary to consider how to prepare problem situations as mentioned above. We were led to adopt open-ended problems for the following reason: When the analysis of a problem situation results in a unique solution, it may happen that (a) the situation involves what students have already learned and (b) it leaves too little room for their preferred way of thinking.

Following these considerations, such problems as "Interim Results of Baseball Games" and "Ranking Teams in a Marathon Race," explained in later chapters, were formulated together with an ordinary test. We tried out these problems with students in several schools at the elementary, lower secondary, and upper secondary levels. We intentionally used the same problem set throughout all school levels, since we reasonably expected that the results of the ordinary test would progressively improve with the progression of grade levels. We believed it would be meaningful to compare this improvement with that of the results using open-ended problems.

We obtained a negative finding with respect to the second question on the survey; that is, the achievement of higher objectives was not necessarily parallel to the achievement that was measured by the ordinary test. We also found that the variation among schools in the achievement of higher objectives was much greater than that in the achievement that was measured by the ordinary test.

During the second phase of our study, we expanded our group in order to replicate the findings of the first phase, as well as to study the next question:

3.   Knowledge, skills, and ways of thinking are important components of the higher objectives, but can these components be further developed by additional instruction? Or does such development depend more on native talent, beyond the influence of teachers? If so, then all that teachers can do is to concentrate on teaching knowledge, skills, and such. Is this really so?

In an attempt to answer the question, experimental teaching was organized and carried out. We formed experimental and control classes, administered a pretest to both groups, and taught the classes by using an open-ended approach two or three times during a three- or four-month period. Teaching in the usual style was done with control classes covering the same topics as the experimental classes. Both groups were given the posttest at the end of the teaching period in order to examine the differences between them.

As usual in this kind of educational experiment, it was difficult to keep the conditions of the experiment strictly constant and the statistical analysis of the data did not lead to a definitive conclusion. However, we arrived at a general conclusion that it is possible to approach the higher objectives through teaching that uses the open-ended approach. The development of students' acquired knowledge and skills as components of higher objectives depends not only on native talent but also significantly on the influence of their teachers in furnishing an opportunity to develop and in advising and encouraging the students. We point out, as a background to this finding, that we assumed that evaluating students' responses to open-ended problems by using the number of different responses and their mathematical quality, as described in later chapters, can be regarded as an evaluation of the higher objectives.

We began to address questions 1 and 2 in 1971, continued our practical study for six more years, and gradually came to the conclusion that introducing this open-ended approach to classroom teaching represents a way of improving mathematics education in Japan. Now we propose to introduce open-ended problems, as exemplified in this book, to ordinary classroom practice that is based on more traditional problems.

THE ORIENTATION OF THE OPEN-ENDED APPROACH IN MATHEMATICS EDUCATION

In the previous section, we proposed to introduce this approach into classroom teaching and stated that such a proposal presupposes that students' achievement of the higher objectives can be evaluated by their responses to open-ended problems. Now we propose to discuss the validity of the assumption and to explain our thinking about the orientation of the student activity that was stimulated by this approach in various stages of mathematical activity. Our discussion is based on the model of student activities shown in figure 1.1.

Mathematical Activities

Many areas of thinking are closely related to mathematics, such as understanding an existing theory of mathematics, solving a mathematical problem, constructing a new theory, or

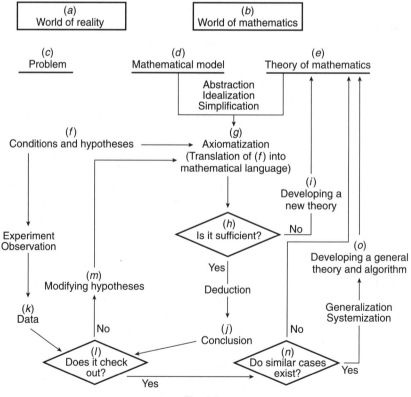

Fig. 1.1
A model of mathematical activities

solving a problem in a nonmathematical field by applying mathematics. We group these areas of thinking together and call them "mathematical activities." At various points in the whole process, we will consider their significance with respect to the relation between the world of mathematics and the world of reality as reflected in the model shown in figure 1.1. This model can be seen to reflect a historical process through which humans have developed today's mathematics as well as a developmental process of a student's learning mathematics in the sense that ontogeny recapitulates phylogeny.

To begin, let us assume that there are two worlds—namely, (a) the world of reality and (b) the world of mathematics—and that there is (c) a problem to solve in the world of reality. Here, (a) may not necessarily be the empirical physical world but may be a conceptual world that is less abstract than (b). For (c), (f) conditions and hypotheses would be formulated from the experiences in (a) and translated into mathematical language through processes of abstraction, idealization, or simplification so that (e) the theory of mathematics might be applied. These processes have much to do with the mathematical theory that the participant in these activities has learned up to that time as well as with the participant's knowledge of (a) because the participant may call on past experience in successfully solving problems of a similar nature, search to find a common feature, and consider its applicability. This stage in which a participant tries to reformulate the problem in his or her preferred way of thinking is called (g) axiomatization.

When mathematics is applied to a real problem, the process always proceeds from (f) to (g), regardless of the participant's awareness of the process. For example, when a teacher counts the number of children after recess in order to confirm that all children are present, the teacher is applying a theory of finite sets and has no need to consider the individuality of each child. Similarly, when a person estimates the work done in terms of workdays, the problem is translated into the world of arithmetic operations by idealizing or simplifying conditions about the additivity of the work and the efficiency of the workers. Another example would be to regard each trial as independent in applying the theory of probability to an actual problem.

A group of propositions formulated at stage (g) needs to be tentatively examined. Do we have enough propositions from which the solution to the problem at hand can be derived? In other words, is it possible to translate any proposition in (a) to one in the axiomatic system (g)? If not, it is necessary to add more propositions to the conditions or hypotheses in (f).

After we have sufficient axioms in (g), we can make a proposition in (g) that corresponds to a proposition in (a). The truth of a proposition in (g) should be decided only by deduction in the axiomatic system (g). For this deduction, (e) is used with the axiomatic system (g).

Even after careful examination, a case may exist in which the deduction cannot proceed as intended. In such a case, it becomes necessary to (i) develop a new theory. For example, students who have learned mathematics up to the lower secondary (junior high school) level may mathematize the movement of a point in space with coordinates that are considered functions of time and may expect that all information about that movement is reduced to the properties of the functions considered. These students, however, do not yet have the theory needed for deduction about the velocity and acceleration of that movement. At this point, they may construct a theory about the derivatives of functions by refining their intuitive idea of velocity. This could be a context in which calculus may be introduced into upper secondary (senior high) school mathematics.

The (j) conclusion, which is derived by deduction, is combined at the stage (l) checking with (k) the data that was obtained empirically in (a). If the two agree within a degree of error allowed in the process of proceeding from (f) to (g), the assumptions will be preserved as tentatively true. If the extent of disagreement is larger than the allowed degree of error, then it becomes necessary to go to (m) modifying hypotheses by regarding some of the assumptions as wrong. In order to assign the reason for the disagreement only to a stage (f), the process from (e) to (j) must have been consistently based on deductive logic. This requirement is an important reason for using demonstration or proof in mathematics.

Now, at the last stage of a process (f) to (g) to (j) to (l), if the result of checking is affirmative, then the axiomatic system (g) is called a mathematical model of (f) and should be examined to determine if any similar case exists. If there is no similar case, the result may be reserved in the participant's (e) theory of mathematics as one typical example of a successful mathematical model. If there are other similar cases, the participant will try to generalize them by studying their common features. The participant will also attempt to systematize the obtained propositions by separating them to more fundamental ones and then to subsidiary and dependent ones. In this way, the participant will proceed to (o) develop a general theory and algorithm in which she or he develops symbolism that is parallel to the general theory just obtained so that the related deductive reasoning can be carried out by transforming strings of written symbols. If we look at only this stage, mathematics appears to be a kind of symbol game. The result of the (o) stage will also be incorporated into (e). One example of such a process is teaching the multiplication of fractions at the elementary school level, where the process begins with concrete problems for purposes of introduction and is finally summarized as an algorithm for the multiplication of fractions that is independent of concrete examples. Another example is the construction of formulas for differentiating functions at the upper secondary school level, which is a summary of step-by-step operations to obtain derivatives of functions.

The (e) theory of mathematics becomes richer and richer through inputs from (i) new theories, (n) similar cases, and (o) general theories. The ultimate source of input is (a), although (b), thus enriched, may play a role of (a) in the sense that its inner integrity or improvement is questioned.

In summary, the process illustrated in the model is a creative activity. For students, in particular, it is a creative process of learning. In classroom teaching, a part of this whole process is taken up piece by piece, according to each stage of a student's development.

Note that our use of the word *model* implies a representation of reality that can be used to arrive at a theory, but the term can have a quite different meaning, as illustrated in the following examples:

1. A *representational model* that is made by giving an interpretation to undefined terms in an abstract axiomatic system. The representation of a group by matrices and the model of non-Euclidean geometry constructed in Euclidean space are examples of this kind of model; however, no room exists for this kind of model in the usual school mathematics curriculum up to the upper secondary level. The purpose of this model is to confirm the consistency of an abstract axiomatic system.

2. *A quasi-mathematical model,* in which the meaning in an abstract theory is associated with words that describe the world of reality. A "fair die" used in textbook exercises in probability is simply an expression of an event with the probability of 1/6 in axiomatic probability theory. If a die is cast in the real world, the probability of 1/6 is an assumption about that die. In exercises where the probability is presupposed to be 1/6, it is another assumption that is adopted for the purpose of making abstract theory more concrete or easier to approach. Similar examples might be found in situations of the so-called application problems or verbal problems used in mathematics teaching. Cars traveling at a constant speed and the unit price of fruit that is constant regardless of the number purchased are examples of such quasi models unless the constant speed and the constant price are actual occurrences. These quasi models are isomorphic to their background abstract concept and have no secondary restriction of applicability in the real world.

It may be said that these quasi models are generated through the process from (n) to (e) and that their use or purpose in teaching is to promote understanding in (e).

3. *Physical models in the empirical world* that are only locally isomorphic to mathematical concepts or relations. A figure drawn on grid paper as a graph of a given function is a physical representation of the abstract concept of the function. It should be used only after understanding its proper convention; further, its representation is local and approximate in nature. It can facilitate our understanding of the function by its appeal to our visual sense and ease in manipulation.

Although solid geometric models and structural teaching aids such as Cuisenaire rods or Dienes blocks cannot have all the properties of geometrical solids or number systems and are only locally isomorphic to what they represent in some aspects, they can nevertheless contribute to students' understanding of concepts because they can be seen and physically manipulated. Even the mathematical symbols referred to previously can be regarded as a model of the general theory in this sense.

Both examples 2 and 3 above are intended to represent abstract concepts or theories in a concrete manner to enhance understanding through easy manipulation of concrete objects.

Orientation of Various Learning Activities

Let us consider traditional learning activities according to the model shown in figure 1.1.

When a teacher introduces a new concept to a class, it is common to begin with introductory problems to help students see a need for the new concept. It is also common to construct the new concept on the students' previous learning and to lead them toward a new theory. If the introductory problems are real-world problems and if teaching starts with translating conditions or hypotheses into mathematical language, the lesson proceeds along the course from (f) to (g) and so on. However, it usually begins from stage (g), in which the translation has been done by someone else (i.e., the teacher or the author of a textbook). Afterward, it proceeds along the course from (g) to (i) to (j) and then to (n) and omits (l). This means that the model used in (g) is actually a quasi model. Next, the usual teaching may proceed to (o), and a set of formal exercises is assigned for assimilation. The activities in the exercises are usually a kind of symbol game.

Furthermore, for assimilation of the new topics, the so-called application or verbal problems are assigned. Although these problems are described in terms of the world of reality, they are usually so well structured that they have one definite answer. For example, in the case of problems about shopping, the cost is assumed to be proportional to the number purchased and "3 dollars each" in a problem actually indicates a special case of a proportional relation between two variables. In the world of reality, this proportionality is a mathematical model agreed to by both sides (buyer and seller) that might be subject to change, as in the example of the purchase of a large number of the item. In the so-called application problems, such a situation would never arise because the shopping situation is used only as a situation in which to apply simple multiplication or division. The situation used in the application problems may then be a quasi model. The course actually taught may not start from (g) as a mathematical model but rather may proceed from (e) to (g') to (j) by substituting the quasi model g' for g. In other words, it aims at the assimilation of a quasi model as a typical example.

A common feature through these stages is that the conclusion in each stage is predetermined, with one logical exception of the generalization from (n) to (o). Even in this exceptional case, however, many teachers use a teaching plan that predetermines the general theory. Other generalizations suggested by students are likely put aside or ignored by the teacher.

In contrast, the processes of abstraction, idealization, or simplification from (f) to (g) and of generalization in its essential sense from (n) to (o) are open ended in such a broader sense that the result is not predetermined (not necessarily in the narrow sense that there are multiple answers, which is what we are using in this book). Students' thinking in these stages involves groping in a repertoire of previously learned skills for a skill, a combination of skills, or a modification of a skill to apply to the present situation. It also involves formulating the situation by selecting the most promising skill through trials. Subsequently, the groping through a deduction following from the formulation and a process from (l) to (m) may be repeated. In this case, the richer the repertoire that the students possess, the more ways of formulation they will have and the higher the quality of formulation that will be available. In this stage students require a high ability to integrate what they have learned. In other words, students need the ability to conceive something new and to change their conception if needed.

The open-ended problems that are proposed here involve the processes from (f) to (g) or from (n) to (o) in a reduced form that permits their use as part of classroom teaching. However, the problems are not of the form that occurs in a real situation, while having a characteristic aspect peculiar to the processes from (f) to (g) or from (n) to (o). This is the basic reason for the assumption mentioned on page 4.

The new proposal in this paper would not be needed if the general objectives of teaching mathematics were concerned only with activities that end in a unique result or answer or if the outcomes of such activities were automatically extended to cover a broader aspect of objectives. In present-day education in Japan, however, the philosophy of mathematics education, supported by many educators and reflected in the general objectives of the present National Course of Study, seems to be concerned with all the stages in the model in figure 1.1. It is empirically denied that the outcomes of learning activities concerned only with arriving at a unique answer would be extended to the whole.

From these considerations, we may reasonably assert that activities corresponding to the processes from (f) to (g) to (h) to (j) to (l) to (m) or from (n) to (o) should be included in an entire program of mathematics education for most students. The proposed activities and those traditionally carried out in the schools should be viewed as complementary to each other, not as alternatives. To promote this complementary relationship, it is preferable to provide students many opportunities to develop solutions from a multiple-aspect point of view by developing small-scale teaching units that use some modification of traditional materials instead of using larger-scale units once or twice throughout the school year that involve all the processes from (f) to (g) to (h) to (j) to (l) to (m) to (g) and so on. One of the main features of this process is that several formulations are possible according to how students conceive of situations in which results are not predetermined. It also affords ample opportunities for them actively to use knowledge, skills, views, and ways of thinking that they have already learned or that come naturally to them.

Our proposed open-ended approach involves developing small-scale problems with few conditions that permit multiple formulations by students and that require no more than the repertoire they have already developed in their education. The next step is to use these problems in a teaching-evaluating program in combination with the other approaches that are already in place in schools.

The explanations in subsequent chapters should be understood as a plan not to replace present practices by this style of teaching but rather to adopt this style as an indispensable part of teaching school mathematics. Most model lessons have already been classroom tested by members of our study group, and to some extent students have reacted as expected. However, among the accompanying similar examples, some that are composed by analogy are yet to be tried.

Furthermore, although we have emphasized for our research purposes problems that elicit plural and genuine results, in practice other types of problems will likely have similar effects if teachers place more emphasis on problems in which the results are not predetermined. We are introducing these exemplary units not because we assert that they are the best among all possibilities but because we believe our main objective can best be promoted through concrete examples. We realize that our exemplary units are subject to improvement.

Chapter 2

An Example of Lesson Development

YOSHIHIKO HASHIMOTO
National Institute for Educational Research

IN THIS CHAPTER we shall explain how the open-ended approach to teaching can be carried out in the classroom. We will use, as an example, the water-flask problem in an experimental lesson that was tried out in classes from the elementary school level through the lower and upper secondary school levels.

THE WATER-FLASK PROBLEM

A transparent flask in the shape of a right rectangular prism is partially filled with water. When the flask is placed on a table and tilted, with one edge of its base being fixed, several geometric shapes of various sizes are formed by the cuboid's faces and the surface of the water. The shapes and sizes may vary according to the degree of tilt or inclination. Try to discover as many invariant relations (rules) concerning these shapes and sizes as possible. Write down all your findings.

The following examples are just a few of the possible answers to the problem:

1. Let a and b represent the lengths of the perpendicular edges from the base to the surface of the water, as in figure 2.1. Then $a + b$ is constant.
2. The midpoint M of segment AC in figure 2.1 is a fixed point, and the segment joining M and the midpoint N of the side opposite AC is a fixed segment.
3. The total area of the sides of the flask under the water surface is invariant.
4. When the flask is tilted as in figure 2.2, $b \times c$ is a constant.
5. The shape of the water surface is a rectangle.

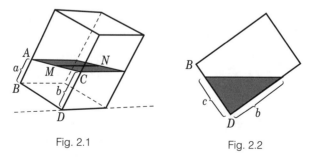

Fig. 2.1   Fig. 2.2

This problem is an open-ended problem as defined in chapter 1; that is, the problem is formulated to yield multiple correct answers.

How will students respond to this problem when it is presented in a typical classroom? To get a rough idea, we carried out the following trial:

> We prepared a set of flasks like that in figure 2.3 for hands-on use by sixth-grade students working in small groups in a classroom in a metropolitan area. We expected that students would better understand the problem situation and be able to discover rules by actually handling the flasks with water in them instead of being presented with the situations on paper. For example, we thought students could discover the rule that "when the water-level readings on the right scale increase by 1 cm, 2 cm, 3 cm, and so on, the water-level readings on the left scale decrease by 1 cm, 2 cm, 3 cm, and so on," or that "there is a fixed point M in plane $ABDC$." We found that students discovered several rules, such as the rule that the sum of $a + b$ is constant (see figure 2.1), rules relating to change and range, and rules concerning the surface of the water. In summary, the trial of this problem demonstrated that students at the sixth-grade level or above dealt with the problem in a meaningful way and with understanding.

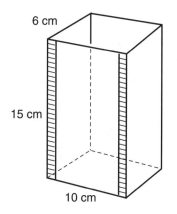

Fig. 2.3
Two scales are graduated by 5 mm;
the lengths show the inside measurements.

When students found the rule that the sum $a + b$ is constant, the teacher then asked them, "Why is that so?" Frequently they replied, "Because we can just see it." Although the students were interested and were enthusiastic about dealing with the situation until they discovered the rule, they seemed not yet mature enough to reflect on why the rule applied.

Active guidance by the teacher was needed to make students aware that the logical relations were also part of the problem. If the teacher's question "Why?" can be transformed to a student's problem through the teacher's guidance, it will then provide a natural situation calling for a logical demonstration by the students. For example, the property of the sum's being constant can be derived without too much difficulty from the fact that the volume of water is constant, as will be seen later.

After various rules have been proposed by the students, the teacher may then encourage them to consider whether all the proposed rules are independent of each other. In doing so, the teacher has an opportunity to help them actually experience an awareness of the fact that from some fundamental propositions others can be logically derived. A discussion of the logic inherent in the rules that are discovered by the students should follow the discovery of the rules; if not, it is difficult to characterize the lesson as sound mathematics teaching, even though finding rules and being able to think diversely are important. We concluded that it was necessary to include such a logical discussion in our program by adjusting the process to the thinking level of the students. We also concluded that this problem of the water flask might also be suitable for use with students in lower or upper secondary schools in addition to the elementary school.

DEVELOPMENT OF A LESSON ON THE WATER-FLASK PROBLEM— A CASE IN THE ELEMENTARY SCHOOL

On the basis of our experience with the first trial, we used the water-flask problem with an open-ended approach at each of three school levels. We describe here a two-hour lesson in an elementary school.

Purpose of the Task and Preparation

1. The task: water-flask problem
2. Subjects: Thirty-nine sixth-grade students in Fukazawa Elementary School in Setagaya Ward, Tokyo
3. Purpose: To have students discover various relations implicit in the problem situation where a flask containing water is tilted; furthermore, if possible, to have students formulate the relations they discover into mathematical expressions and to explain logically the formulated relations
4. Instruments and materials:
   a) Ten water flasks (as in fig. 2.3)
   b) Ten beakers for pouring water
   c) A blank sheet of paper for each student
   d) Ten blank transparency sheets for presentations by groups on the overhead projector

   (*Note*: We used a science laboratory because it was more convenient than an ordinary classroom for the use of beakers, flasks, and water and for discussion.)

Sequence of Presentation and Allocation of Time

The teacher used two forty-minute periods for the lesson. Two periods were used because other study-group members had found from their experience that more than one period was necessary. In most instances, one period allows insufficient time for discussion and a summary of all the students' findings, many of which are very interesting. Two periods, however, provided enough time for students to explore the problem situation and discuss most, if not all, of their findings.

In general, when teaching with the open-ended approach, the teacher must carefully allocate and manage time because students are likely to generate many responses, both expected and unexpected, and all should be discussed and summarized. For example, in chapter 4, one teacher conducting this lesson mentioned that "one point to be reconsidered is that it took too much time (60 minutes). It would be ideal if I could allocate two periods, not one as I did, and spend the first hour for finding methods and the second hour for discussion." Accordingly, the flow of teaching for this lesson was planned as follows:

The first period: Individual work follows the presentation of the problem to the whole class. Each student is given a blank worksheet on which to write his or her ideas. The worksheets are collected afterward by the teacher for use in preparing a summary of individual students' responses. In the group work that follows, each group of four students uses one set of instruments and a representative of each group writes down the results of the group's discussion and also, if possible, the process by which the group arrived at the result.

The second period: The results of each group's work are presented and discussed. Then the lesson is summarized.

The Lesson Plan

The First Period

Teacher's Presentation and Directions	Students' Activities	Remarks	Cumulative Time in Minutes
1. When we tilt the flask with water in it while fixing one edge of the base on the table, we see that the shape and size of various parts are changing. Find out as many relations among the parts as possible, and write them down.	1. Understanding the question	1. Explain the problem by using a real flask with water. 2. Use figure 2.4 as a poster to make sure the students understand the problem. Fig. 2.4	5
2. Write down what you have noticed on the blank worksheet.	2. Trying to find various rules (individual work)	3. Distribute sheets to each student. 4. Collect sheets on which students have written their findings.	25
3. Within each group, discuss what you have found. The leader of each group should record the group's observations.	3. Discussing within groups, and discovering various rules (group work)	5. Distribute new sheets to each group.	40

The Lesson Plan (continued)

The Second Period

Teacher's Directions, Questions, and Activity	Students' Activities	Remarks	Cumulative Time in Minutes
1. Please present the results of your group discussion.	1. Groups take turns presenting their results.	1. List every response from the groups on a poster.	20
2. Let's group together similar findings.	2. Rules are grouped from various viewpoints.	2. Have students group findings carefully so as not to duplicate or omit any.	30
3. We know the rule that $a + b$ is constant, where a and b are the lengths of the sides shown in 2.4. Can we explain the figure rule?	3. Students methodically consider why the property of the sum's being constant is true.	3. Assign a and b to the bases of the shaded trapezoid as in figure 2.4.	40
4. The teacher gives the reason, if necessary.	4. Students listen to the explanation.		
5. Can we put in order other rules from different points of view?	5. Students summarize their findings.		

Description of the Actual Lesson and Related Discussion

1. *Time allocation.* The teaching proceeded roughly as planned, but only five of the ten groups had time to present their results fully. The remaining five groups presented their results on the overhead projector, and discussions were limited to those findings to which other groups had not referred.

2. *Individual work.* The teacher walked around observing the students' activity as they worked at their desks. Students were able to write several rules relating to the constant sum, variation, range, and water surface. Specific guidance was generally not necessary.

3. *Group work.* Most groups worked successfully and were able to summarize their findings. While walking around, the teacher gave the following advice to two groups that had noticed that the amount of increase on one side was just equal to the decrease on the other: "You have found a good point, but isn't it possible to say it in another way? Think a little more." The two groups found that the sum was constant. (They did so by observation, not by making a table.) The teacher asked the five groups that found that the sum was constant for a reason for the rule. No group was able to give a correct response. So the teacher encouraged students to continue trying to determine why the sum was constant.

4. *Group presentations.* Students were active, asking questions and giving answers, during group presentations. One factor that gave rise to this activity was that the results discussed within a group were better and more firmly summarized than those arrived at individually.

The following dialogues are examples of such discussions. (*T* represents the teacher and *S* the student; notes by the author are given in parentheses.)

Example 1. The rule that the sum is constant is developed.

S1: The sum of the scale readings on both sides is the same when the prism is tilted because the area of this part (trapezoid) doesn't change.

S2: Just a question. Can I ask?

T: OK.

S2: What S1 says is that when this part is scaled as 1, 2, 3, 4, and so on, and this is 6 and this is 2, then this sum and that sum are the same. [4 + 4 = 6 + 2] Isn't it? But what will happen when this side is cut like this? (In a triangle, as shown in fig. 2.5)

The discussion proceeded to the next step, whereas discussion about the question raised by *S2* was reserved. In responding to S1's presentation, *T* asked, "Why doesn't this area change?" *S1* answered, "If we draw a line through this (the fixed point in the middle viewed from the front), just the decreased amount in the left side appears in the right side, and so the area doesn't change." In other words, the student did not consider an explanation based on the fact that the total volume of water is constant.

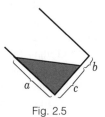

Fig. 2.5

Next, in the presentation by *S2*'s group, the representative (not *S2*) gave the following rule:

S3: If this *b* is subtracted from length *a*, the difference is always the same [$a - b$ is constant in fig. 2.5].

T: Is this true?

S3: Here we regard the upper part from this as plus, and in the other as minus. When it is minus, then it will be subtracted from one side.

T: It is OK? ("I cannot follow" was heard from all the other students.)

T: Now, we will reserve it for later discussion.

An interesting point was noticed, but the conclusion was false. In this example, the product of *a* and *c* is constant. After the lesson, the teacher made a comment on this relation to S2's and S3's group.

The next example is concerned with the presentation by another group.

Example 2. The total surface area does not change.

S4: What do not change are the volume and the surface area.

S5: I have a question.

T: Yes, please, *S5*.

S5: As the area of the water surface (upper face) changes, the total surface area also changes, doesn't it?

T: How about *S5*'s comment?

S6: I beg your pardon.

T: What he said is that as the area of this, the area looked down on from above, changes, the surface area must change accordingly. (The students' comments "absolutely change" or "may change" were heard.)

The group to which *S5* belonged had written these three rules regarding the area in its report: (1) The area of the water surface changes, (2) the total surface area changes, and (3) the total surface area except that of the water surface does not change. In short, since the students had already found these rules, they seemed able to make an objection to *S4*'s presentation. At that stage, students could summarize that what was constant was not the total surface area but rather the total surface area of the sides.

All the results of the students' activities in both stage 2 (individual work) and stage 3 (group work) are summarized in table 2.1.

5.   *Grouping rules by similar content.* The rules that students found are grouped under the following headings: (1) constant sum, (2) variation, (3) range, (4) shape of water surface, (5) area, (6) volume, and (7) others. Table 2.1 summarizes the results of the students' group work and shows that, of ten groups, seven found that the sum was constant and two found that one side decreased by the amount the other side increased. With these results, further development of the lesson became easier. (*Remark*: If many students fail to notice a rule referred to above, a good way to help them is to let them draw a line segment (e.g., using a water-soluble color pen) along the water level on the side (or face) of the flask.)

Actually, in another experimental class, while walking around observing students work at their desks, the teacher observed that many groups did not notice the property of the constant sum. The teacher encouraged them to draw a line segment as mentioned above and then found that students were able to see the property of the constant sum or that one edge decreased exactly by the amount that the other increased.

6. *The logical explanation.* We intended, as mentioned earlier, that the logical explanation for the rule that $a + b$ is constant would be included in the lesson only when the teacher judged it possible. The teacher observed the students as they worked at their tables during the lesson and found their activities and group presentations to be active and rich. The teacher then decided to take up the problem of the logical explanation.

Subsequently, the teacher led all the students to confirm the property that $a + b$ is constant by letting them make a table, as shown in figure 2.6, in which they could enter the data they collected. As the next step, the teacher asked, "Can anyone explain why $a + b$ is constant?" The teacher found that just one student could give the reasoning, whereas the others seemed unable to grasp the question immediately. The teacher then brought out a large solid model made of cardboard as shown in figure 2.7 and explained the reason in the following way: The volume is constant. When we regard the solid as a prism whose base is a trapezoid, the height of the prism is constant, the area of the trapezoid is constant, and its upper base plus its lower base is constant. Therefore, $a + b$ is constant.

Table 2.1

Summary of Individual and Group Observations of Thirty-nine Students

Categories of Observations (Findings)	Number of Rule	Students' Observations (Rules)	Group Numbers										Number of Students Making Observation
			1	2	3	4	5	6	7	8	9	10	
Constant sum	1	$a + b$ is constant.	*	*	*		*			*		*	6
	2	The sum of the lengths of the edges above the water surface is constant.				*							1
Variation	3	One edge decreases by the amount the other increases.						*	*				4
	4	When one edge increases, the other decreases.											2
	5	The lengths of the edges vary.				*	*						5
	6	The length of the edge of the water surface becomes greater.				*			*				0
	7	When one edge becomes 0, the other edge becomes twice its original length.											1
Range	8	The limit of the length of an edge is 15 cm.								*	*	*	9
Shape of water surface	9	The water surface (upper) and the base are rectangles.	*										0
	10	The water surface is a rectangle or a quadrangle.					*				*		3
	11	The shape of the base is constant.						*				*	4
	12	The shape of the side plane changes from trapezoid to triangle.							*		*	*	7
	13	The side view is a trapezoid.											1
	14	The shape of the water surface changes.						*					9
	15	In some instances, the base becomes smaller.									*		4
Area	16	The total area of the side faces does not change.					*	*	*		*		3
	17	The area of the water surface changes.	*										9
	18	The area of the water surface becomes larger.		*							*		5
Area	19	The area, except that of the water surface, does not change.	*										1

Table 2.1 (continued)

Summary of Individual and Group Observations of Thirty-nine Students

Categories of Observations (Findings)	Number of Rule	Students' Observations (Rules)	Group Numbers										Number of Students Making Observation
			1	2	3	4	5	6	7	8	9	10	
Area	20	The total surface area changes.	*										4
	21	The area of the base is smaller than that of the upper face when tilted.											1
Volume	22	The volume does not change.	*	*		*	*	*	*	*	*	*	13
	23	When the water forms a prism, the volume equals (base area) × (height).											1
Others	24	There is a fixed point, when viewed horizontally.	*	*									1
	25	The weight of the water does not change.	*								*		0
	26	The angles change.					*				*	*	6
	27	The sum of the angles of the side planes is constant.					*						0
	28	The water surface is level.				*			*				
	29	The form of the water is a quadrangular prism.						*					1
	30	The form of the water changes from a cuboid to a triangular prism.											2
	31	The form of the water changes.									*		5
	32	Others									*	*	4
Total number of rules observed, by group			8	4	2	5	7	7	3	4	10	8	114

Note 1: The rightmost column shows the number of students who observed the rule during individual work. For example, six of thirty-nine students observed the rule that $a + b$ is constant.

Note 2: An asterisk indicates that the rule was observed by a group during work.

Note 3: The numbers in the last row indicate the total number of rules observed by each of the ten groups. For example, group 1 succeeded in observing a total of eight rules: 1, 9, 17, 19, 20, 22, 24, and 25.

a			
b			
$a + b$			

Fig. 2.6
A table for entering students' data

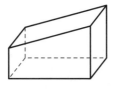

Fig. 2.7
A solid model used to explain that $a + b$ is a constant

We should comment that it might have been possible to use the water flask in the explanation, but in this instance, when the flask is held so that its trapezoidal base is horizontal, the water flows out of the flask. This cardboard model helped students to understand the reasoning easily.

VARIATIONS OF THE WATER-FLASK PROBLEM

We have presented one lesson plan used at the elementary school level. Many other ways exist to organize the lesson. For example, we could expand the lesson to a group of similar problems.

One Side of the Flask Is Not Fixed

In the previous section, the problem was presented with the condition that the position of one edge of the water flask's base was fixed on the table (plane). At a higher grade level, however, it would be possible to extend and present the problem without such a restriction.

In this instance, the lengths of the four sides (edges of the flask) below the water level may change according to how the flask is tilted. When the students are led to consider how to classify the flask's inclination and to find the two cases (1) where just one *point* is fixed (i.e., in a "free" position) and (2) where one *side* is fixed, they may notice that the relation is easier to formulate in the latter case than in the former, since it can be reduced to a problem involving two variables by grouping four variables into two pairs of equal length. Students can first study the case with the fixed side and then extend the problem to the case in which one point is fixed.

Accordingly, let $ABCD$ and $A'B'C'D'$ represent the base and the water surface, respectively, and $AA' = a$, $BB' = b$, $CC' = c$, and $DD' = d$ as shown in figure 2.8. Point D is fixed. Given the original water-flask problem, at least these four observations emerge:

1. The quadrilateral $A'B'C'D'$ is a parallelogram.

2. The point O', at which $A'C'$ and $B'D'$ intersect, is a fixed point.

3. $a + b + c + d$ is a constant.

4. $a + c = b + d =$ a constant.

These relations can be obtained as an extension of those in the case in which one side is fixed on a tabletop. (*Note*: In figure 2.8, let O be the point of intersection of the diagonals of the base $ABCD$ and O' be the intersection of the diagonals of the water surface $A'B'C'D'$; then $\overline{OO'}$ is the median of the trapezoid $BB'D'D$. Since the quadrilateral $BB'D'D$ is a trapezoid, $2(OO') = b + d$. Similarly, $2(OO') = a + c$, from which the first half of observation 4 follows (1).

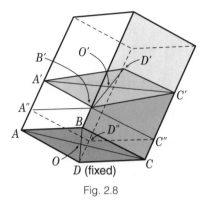

Fig. 2.8

Let V be the volume of the polyhedron $ABCD\text{-}A'B'C'D'$, and V_1, V_2, and V_3 be, respectively, the volumes of three polyhedrons that are produced by cutting $ABCD\text{-}A'B'C'D'$ by a plane parallel to the base through B', that is, the rectangular prism $ABCD\text{-}A''B'C''D''$ and the quadrangular pyramids $B'\text{-}C'D'D''C''$ and $B'\text{-}D'D''A''A'$. Let $AB = p$, $AD = m$, and $S = pm$.

Then we have

$$V = V_1 + V_2 + V_3,$$

$$V_1 = pmb = bS,$$

$$V_2 = \frac{1}{3} \times (C'C'' + D'D'') \times \frac{p}{2} \times m = \frac{1}{6} \times (c + d - 2b)S,$$

$$V_3 = \frac{1}{3} \times (A'A'' + D'D'') \times \frac{m}{2} \times p = \frac{1}{6} \times (a + d - 2b)S.$$

Therefore,

$$V = \frac{1}{6} \times (6b + c + d - 2b + a + d - 2b)S$$

$$= \frac{1}{6} \times (a + c + 2b + 2d)S.$$

Since V and S are both constant regardless of the position of the water flask, the second half of observations 4 and 3 follow from this formula and (1). Accordingly, observation 2 follows from the fact that OO' is constant.

The Flask Has a Triangular Base (for Upper Secondary School)

It is also possible to extend the problem by changing the shape of the base of the water flask. In this instance, a new problem can be formulated as follows:

Suppose that we have a water flask in the form of a triangular prism that is half full. The flask is tilted while one side of the base is fixed on a tabletop. Many quantitative or geometric relations involving various parts of the flask are implicit in this situation. Try to discover as many of them as possible and give the reasons why such relations hold.

The following is a description of a lesson tried in two classes of tenth-grade students in a metropolitan upper secondary school, each consisting of nine groups of four students and two groups of five students. The lesson was carried out using triangular-prism water flasks whose bases were equilateral triangles (10 cm on a side) and whose heights were 15 cm, with no graduations. These dimensions were selected for purposes of easy construction. One classroom period of fifty minutes was used in the lesson.

Students' activities

The teacher gives a water flask to each student, along with a worksheet on which figure 2.9 is printed. (The midpoints M and N, of $A'B'$ and $A'C'$, respectively, and x and y were not printed on the students' worksheets.)

1. *Individual work.* Students try to discover relations involving variables and constants after pouring water of a suitable volume into the water flask and tilting it at various angles.

2. *Group work.* Each group of students summarizes its findings on a worksheet, along with the reasons why the findings are true. All results are written on the worksheet.

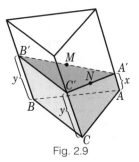

Fig. 2.9

3. *Each group presents its summary.* The findings of all the groups are listed below. The numbers in parentheses represent the number of groups reporting the finding in classes A and B, respectively.

1. $x + 2y = k$ ($AA' + BB' + CC'$ is constant).   (4, 5)
2. The total area of the side planes under the water level is constant.   (2, 4)
3. The water surface passes through a fixed point.   (1, 3)
4. The shape of the water surface is always a triangle.   (3, 6)
5. Determining the range of variables   (2, 2)
6. Determining the center of gravity   (3, 0)

4. *Other findings.* The following relations were also noted:

7. The areas of the two side faces are equal.
8. $B'C'$ is parallel and equal to BC.
9. The water surface is parallel to the top of the table.
10. Each half of the decrease in AA' is added to BB' and CC', respectively.
11. $C'A$ and $B'B$, $B'A$ and $C'C$, and $B'C'$ and AA' are in a skew position to each other.

In addition to those above, other findings were made (e.g., the length of MN is constant and the segment MN is fixed), though these were not discussed in the lesson.

The lesson did not progress to the stage where the proof for $x + 2y$ being constant was considered by all students. Therefore, another lesson was carried out with a class of tenth-grade students at another school in order to derive the proof through group work. In this lesson, actual flasks were not used and students were provided only with blank worksheets (no figure) on which to write their proofs. In instances where groups did not successfully derive proofs, the main source of difficulty was found to be incorrect drawings or incorrect manipulation of algebraic expressions.

The following proof was given by one of the successful groups (see fig. 2.10):

Let h be the height of the base triangle with side p. The volume of water is

$$\frac{(a - b)ph}{3} + \frac{bph}{2}.$$

Therefore, $\frac{a - b}{3} + \frac{b}{2}$ is constant and $\frac{2a - 2b + 3b}{6} = \frac{2a + b}{6}$.

Therefore, $2a + b$ is constant.

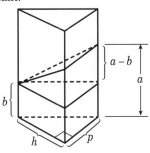

Fig. 2.10

CONCLUSION

Following these experimental trials of the water-flask problem, we concluded that the problem has great potential for use both with lower secondary school students who are interested in mathematics and with ordinary upper secondary school students. We recommend that instruction begin with one side of the base of the rectangular prism fixed and then gradually proceed to change the conditions of the problem by (1) changing the position of the water flask and (2) changing the shape of the base (see fig. 2.11).

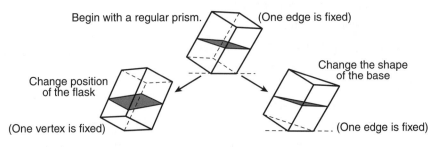

Fig. 2.11

Chapter 3

Developing Lesson Plans

Toshio Sawada
National Institute for Educational Research

IN THE OPEN-ENDED APPROACH the teacher gives the students a problem situation in which the solutions or answers are not necessarily determined in only one way. The teacher then makes use of the diversity of approaches to the problem in order to give students experiences in finding or discovering new things by combining all the knowledge, skills, and mathematical ways of thinking they have previously learned.

Classroom activities are structured to help students—

- mathematize situations appropriately;
- find mathematical rules or relations by making good use of their knowledge and skills;
- solve the problems;
- check the results;

while

- seeing other students' discoveries or methods;
- comparing and examining the different ideas;
- modifying and further developing their own ideas accordingly.

We have been using the open-ended approach in elementary, junior high, and senior high schools for many years. Examples given in chapter 2 and chapters 4–6 are the result of such experiences. We can summarize the advantages and disadvantages of this teaching style as follows.

ADVANTAGES AND DISADVANTAGES OF THE OPEN-ENDED APPROACH

Advantages

1. Students participate more actively in the lesson and express their ideas more frequently.
2. Students have more opportunities to make comprehensive use of their mathematical knowledge and skills.

3. Even low-achieving students can respond to the problem in some significant ways of their own.

4. Students are intrinsically motivated to give proofs.

5. Students have rich experiences in the pleasure of discovery and receive the approval of fellow students.

Disadvantages

1. It is difficult to make or prepare meaningful mathematical problem situations.

2. It is difficult for teachers to pose problems successfully. Sometimes students have difficulty understanding how to respond and give answers that are not mathematically significant.

3. Some students with higher ability may experience anxiety about their answers.

4. Students may feel that their learning is unsatisfactory because of their difficulty in summarizing clearly.

Though some disadvantages exist in using this teaching style, we think we can resolve them. In the following sections, we present some remarks on developing a lesson plan—including how to make a suitable problem, how to use the problem in classroom teaching, and how to evaluate students' activities—in a way that retains the teaching style's advantages while resolving its disadvantages.

EXAMPLES OF PROBLEMS AND THEIR CLASSIFICATION

Let us consider some problems in which some data are given and then find some valid general rules or propositions. By data we mean not only numerical data but any information given to the students. The number of propositions should be appropriate for the level of the students' abilities. For example, when students are asked to discover as many number patterns in Pascal's triangle as they can, they may list many that are not necessarily independent and that can be reduced by students at a more advanced level to only two propositions: one with the number 1 at both extremes of each line and the other the defining recurrence formula. In addition to the water-flask problem discussed in chapter 2, we present some other examples of problems.

Example 1. Baseball standings

The context of the problem in figure 3.1 can be found in newspapers and is familiar to students. Also, students know that baseball games can end in a draw or tie in Japan. Students are asked to find some general rules or relations not only specific to this situation but also valid for others. Giving another such example might help students find various general rules or relations.

> The following table shows the record of five baseball teams. Certain regularities (rules or relations) can be found among the numbers in the tables. Write out as many of these as you can.

Team	Games	Wins	Losses	Draws	Winning Ratio	Games Behind
A	25	16	7	2	0.696	—
B	21	11	8	2	0.579	3.0
C	22	9	9	4	0.500	1.5
D	22	8	13	1	0.381	2.5
E	22	6	13	3	0.316	1.0

Fig. 3.1

Many relations can be observed in the table in figure 3.1; among them are the following:

1. The additive relation among the number of games, wins, losses, and draws:
   (Number of games) = (number of wins) + (number of losses) + (number of draws)
2. The multiplicative relation among the winning ratio, number of wins, and number of losses:
   (Winning ratio) = (number of wins) ÷ [(number of games) – (number of draws)]
3. The number of games behind is determined by the relation between the numbers of wins and losses of two teams.
4. The total number of games is even.
5. The total number of wins is equal to the total number of losses.

As mentioned, the purpose of this problem is to have students find as many rules or relations as possible from several points of view. The rules may range from lower to higher levels; for example, rule 1 is easy to find, whereas others are more difficult, since they involve somewhat more complicated processes.

Example 2. The marble problem

In the problem in figure 3.2, the students are asked to quantify the degree of scattering. Generally, no unique working interpretation has been devised for two-dimensional scatterings, so several ways of quantification are possible according to a variety of points of view. The problem can be used in both the elementary and the secondary schools.

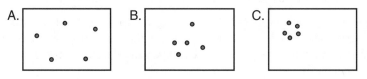

Three students, A, B, and C, each threw five marbles, which came to rest as shown. In this game, the winner is the student with the smallest scattering of marbles. The degree of scattering seems to decrease in the order A, B, C. Devise as many ways as you can to express numerically the degree of scattering.

Fig. 3.2

Students may discover the following methods for measuring the scattering:

1. Measure the area of a polygonal figure.
2. Measure the perimeter of a polygonal figure.
3. Measure the length of the longest segment connecting two points.
4. Sum the lengths of all segments connecting two points.
5. Sum the lengths of the segments connecting one fixed point with all other points.
6. Measure the radius of the smallest circle including all points.
7. Calculate the standard or average deviation using a coordinate system.

Each method has advantages and disadvantages. When the problem is posed to students, some may make a polygonal figure by connecting points and then try to find its area. Other students, however, may criticize this method because, for example, if all points are on a straight line, such an approach will lead to difficulty. In such an instance, it is important for the teacher to help students see both the advantages and disadvantages in generalizing the proposed method of measuring.

Example 3 is formulated by modifying the following problem, which is taken from a sixth-grade textbook:

Several solid figures are shown here:

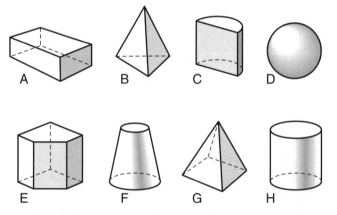

Which figure is a pyramid? Which figure is a cylinder? etc.

For this problem, students are asked to select figures according to different fixed attributes. This exercise offers students practice in identifying figures by using their previous knowledge. We reformulated this problem into an open-ended one that may stimulate more thinking (see example 3).

Example 3. Classifying several solid figures (Refer to "Classification of Solid Figures" in chapter 4.)

Depending on the student's approach to the problem in figure 3.3, groups of figures can be formed according to various characteristics. Students are free to choose their method of classification. This problem can be used to summarize what they have learned in classroom activities. It can also be used as an introductory topic to develop a future lesson plan. Students might produce original ideas in the process of solving the problem, and making use of those ideas in subsequent teaching sessions would be an excellent way to increase their motivation to study.

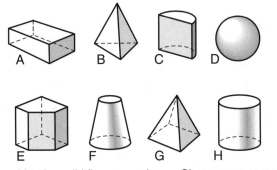

Consider the solid figures as shown. Choose one or more figures that share the same characteristics with figure B and write down those characteristics. Next, choose one or more figures that share characteristics with figure H and write down those characteristics.

Fig. 3.3

Example 4 is an open-ended problem that was reformulated from an ordinary exercise used in the teaching of functions at the lower secondary school level.

Example 4. Finding some common properties

In figure 3.4, students may be able to discover a variety of characteristics, such as the rate of change, algebraic expressions, the shape of a graph, the domain of the function, and so on. Using this problem in classroom teaching may enable students to integrate what they have learned about the concept of linear function.

The open-ended problems presented above serve several purposes, such as developing students' thinking ability and helping them to think from different points of view. The problems also include several examples of mathematical thinking, some of which are elementary and some more advanced. The problems have mathematical value, and they are capable of extension. We call a problem that has these characteristics a "good" open-ended problem for use in mathematics teaching.

Types of Problems

From the examples above, we may classify these open-ended problems into three types:

Type 1. Finding relations. Students are asked to find some mathematical rules or relations. See example 1 (baseball standings) and example 4 (finding some common properties).

Type 2. Classifying. Students are asked to classify according to different characteristics, which may lead them to formulate some mathematical concepts. See example 3 (classifying several solid figures).

Type 3. Measuring. Students are asked to assign a numerical measure to a certain phenomenon. Problems of this kind involve several facets of mathematical thinking. Students

Group A includes a graph and a number table. Group B includes some algebraic expressions that represent functions.

Group A (1)   (2)

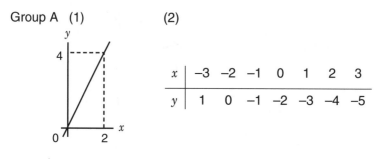

x	−3	−2	−1	0	1	2	3
y	1	0	−1	−2	−3	−4	−5

Group B

(a) $y = (2/3)x$ (b) $y = -x$ (c) $y = 2x + 1$ (b) $y = x^2$

(e) $y = 1/x$ (f) $y = x + 2$ (g) $y = (1/2)x - 2$

Examine the graph (1) and the table (2) in group A and choose functions from group B that share a common characteristic with (1) and (2) according to your classification. Explain your decision. Find as many common characteristics as you can.

Fig. 3.4

are expected to apply mathematical knowledge and skills they have previously learned in order to solve the problems. See example 2 (the marble problem).

Open-ended problems of these types involve much mathematical content and are also effective as applications of mathematics.

HOW TO CONSTRUCT A PROBLEM

Generally speaking, it is difficult to develop good, appropriate, open-ended problems for students at different grade levels. Through repeated trial and error in our research, however, we have derived the following guidelines for creating such problems:

1. Prepare a physical situation involving some variable quantities in which mathematical relations can be observed.

- *Example 1*: The water-flask problem (see chapter 2)

- *Example 2*: Present a tape recorder that is recording. Observe some quantities varying with time and discover mathematical relations among them.

2. Instead of asking students to prove a geometry theorem like "if P, then Q," change this problem to "if P, then what kind of relationship among the elements in the figure can you find?" (*Note*: Several elements may be specified.)

- *Example 3*: Consider the following geometry proof:

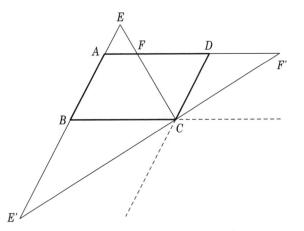

Prove: If figure *ABCD* is a parallelogram and ray *CE* is the bisector of angle *BCD*, then *AE* = | *AD* – *AB* |.

This problem can be modified to the following open-ended one:

> If figure *ABCD* is a parallelogram, ray *CE* is the bisector of angle *BCD*, and *E'F'* is the bisector of the exterior angles of parallelogram *ABCD* at *C*, then what relations can be found among the segments, angles, and triangles?

It is necessary for the teacher to help students understand the meaning of "relations among geometrical figures" in daily lessons before this kind of open-ended problem can be used in the classroom. For example, when considering relations among segments, students should be able to recall such relations as equality, inequality, ratio, parallelism, perpendicularity, and such.

3. Show students some geometric figures that concern a geometry theorem. Then have them draw other figures like the given ones and ask them to conjecture a theorem that is suggested by the figures.

- *Example 4*: Show students the following figures:

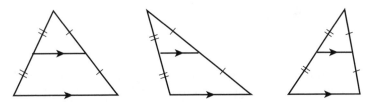

These figures relate to the theorem concerning the segment joining the midpoints of two sides of a triangle. We assume that the students are familiar with the marks used in the figures that indicate relations. A theorem and its converse can be conjectured at the same time.

4. Show students a number sequence or number table and then ask them to discover some mathematical rules.

- *Example 5.* Find as many number patterns as possible in the following table.

1	2	3	4	5	6	7	...
6	9	12	15	18	21	24	...
27	36	45	54	63	72	81	...
108	135	162	189	216	243	270	...

This table is produced by the recurrence formula
$f(m, n) + f(m, n + 1) + f(m, n + 2) = f(m + 1, n), f(1, n) = n$,
where $f(m, n)$ is a term in the mth row and the nth column.

It can be observed, for example, that progressions taken along the left-down diagonal are geometric with a common ratio of 3 or that a progression in the mth row is an arithmetic one with a common difference 3^{m-1}, and so on. Proving these rules from the defining recurrence formula is an exercise in applying mathematical induction.

5. Show students several concrete instances in several categories. Point to one of the instances as an example and ask students to enumerate others that have the same characteristics as the example. Multiple methods of enumeration can be based on a multitude of ways to characterize the examples.

- *Example 6.* Show students several geometric designs that are based on tessellation patterns in wallpaper, and point to one of the patterns as an example. The characteristics should be considered in terms of the direction of translation, symmetry, or rotation. The process to arrive at a conclusion is similar to that of concept formation in that a set of objects is formed by identifying a common property among them.

6. Show the students a group of several similar exercises or problems. Ask the students to solve them and then to find as many common properties as possible among at least two of them.

- *Example 7.* 1. Graph the following functions.
  2. Write down as many properties as possible that two or more of the following functions have in common.

  $(a)\ y = \frac{2}{5}x$ $(b)\ y = \frac{2}{5}x^2$ $(c)\ y = \frac{2}{5}x^3$

  $(d)\ y = -\frac{2}{5}x$ $(e)\ y = -\frac{2}{5}x^2$ $(f)\ y = -\frac{2}{5}x^3$

Originally, these exercises were of the closed type, with such questions as "Which is an increasing function?" "Which is a function whose domain is a proper subset of real numbers?" and so on. Modifying them in this way makes it possible for students to think from their own viewpoints and, accordingly, to be free in their responses.

7. Show the students several quasi-mathematical situations in which a certain difference can be observed. Ask them to find methods to measure the difference.

- *Example 8.* Several persons rank their preference for several drinks. Devise a measure to determine how closely their rank orders agree.

Many ways of measuring are possible, such as the rank-order correlation of Spearman or Kendall and the lexicographic distance or scoring methods, depending on the approach to determining the degree of agreement in rank order.

- *Example 9*. The marble problem given earlier

8. Show the students a concrete example for which an algebraic structure exists (e.g., group structures) and numerical data are easily collected. Then ask students to find mathematical rules that seem to be true.

- *Example 10*. Prepare two paper disks of different sizes, called A and B, where the radius of A is larger than that of B. Mark 0, 1, 2, 3, 4 on the edge of each of the disks in equal distances. Join the disks at their centers by an eyelet so they can be freely rotated around the center (like a circular slide rule). Now, a new addition, symbolized by $\oplus$, is defined using the two disks as follows: The addition $a \oplus b$ means the reading a on the A disk that lines up with b on the B disk after rotating. So, when 0 on the B disk lines up with 2 on the A disk, the reading on the A disk that lines up with 4 on the B disk is 1; that is, $2 \oplus 4 = 1$. Similarly, $3 \oplus 2 = 0$ and $3 \oplus 4 = 2$. What rules are true with respect to this addition?

This example embodies the additive group of residues modulus 5. Students are expected to discover several properties of this group experimentally, though it is left to a more advanced level to list all the axioms of this group.

We can draw the following conclusion about the relationships among the eight guidelines above and the types of problems previously mentioned: If a problem of type 1 (finding relations) is desired, then choose strategies 1, 2, 3, 4, and 8; for a problem of type 2 (classifying), choose strategies 5 and 6; for a problem of type 3 (measuring), choose strategy 7.

HOW TO DEVELOP A TEACHING PLAN

Determine If the Problem Is Appropriate

After the teacher has developed an open-ended problem according to the guidelines mentioned above, it is useful to consider the following three points before using the problem in classroom teaching.

Is the problem rich in mathematical content and valuable mathematically?

The problem should encourage students to think from different viewpoints. However, this alone is not sufficient, for it also should be so rich in mathematical content that both high-achieving and low-achieving students can solve it using several different approaches, each of which has mathematical value.

Is the mathematical level of the problem appropriate for the students?

When students solve an open-ended problem, they need to use previously learned mathematical knowledge and skills. If the teacher judges a problem to be beyond the students' ability, then the problem should either not be used for classroom teaching or be revised. In general, the difficulty of the problem should be appropriate for the students' ability; however, if the teacher wants to use an open-ended problem to evaluate, it is not necessary that proving the general proposition to be conjectured be within the range of students' ability. The students' conclusion that such and such seems to be true in a situation would be significant in itself, even though the students may not be able to give the reasons.

Does the problem include some mathematical features that lead to further mathematical development?

Among the possible responses from students to an open-ended problem, some should be connected or related to some higher mathematical concepts or be capable of further development to a higher level of mathematical thinking.

Develop the Lesson Plan

Assuming that an appropriate open-ended problem is obtained as mentioned above, the next step is to develop a good lesson plan. For this step, the teacher should consider the following points.

List the students' expected reponses to the problem.

Students are expected to respond to the open-ended problem in different ways. Accordingly, the teacher should write a list of their anticipated responses to the problem. Since the students' ability to express their ideas or thinking may be limited, they may not adequately verbalize or explain their problem-solving activities. They may also explain the same mathematical idea in different ways. It is important that the teacher list as many as possible of the students' responses in their own language, even though the responses may be reducible to fewer general propositions. In addition, the list should include more higher-level responses than may be expected for the level of the students.

Afterward, these responses should be rearranged and grouped according to the viewpoints involved and summarized into a general proposition for each viewpoint. For each response, the teacher should clarify the intrinsic mathematical value or a direction for further development.

Make the purpose of using the problem clear.

The teacher should understand the role of the problem in the whole lesson plan. The problem can be treated as an independent topic, as an introduction to a new concept, or as a summary of the students' learning. From our experience, an open-ended problem is especially effective when used as an introduction or as a summary.

Devise a method of posing the problem so that students can easily understand the meaning in the problem or what is expected of them.

The problem has to be expressed so that students can easily understand it and find an approach to solve it. In some instances, they may get confused when the teacher's explanation of the problem is too brief. Such confusion may result because the teacher wants to give students ample freedom to approach the problem or because they have little or no experience in learning other than following the textbook. To avoid confusion, the teacher should pay close attention to how the problem is posed or presented.

Make the problem as attractive as possible.

The problem should be concrete and familiar to students. It should also include aspects that arouse their intellectual curiosity. Since solving an open-ended problem requires time to ponder and think, the problem should be attractive enough to hold students' interest. For the aforementioned example 8, several ways of preparing a concrete situation are possible; a good example is the problem of the beauty contest (see chapter 6).

Allow enough time to explore the problem fully.

Sometimes more time than expected is required to pose a problem, have students solve it, discuss the approaches and solutions, and summarize what has been learned. Accordingly, the teacher needs to allow enough time to explore the problem. Sufficient time should especially be allocated for discussion. Active discussion among the students and between the students and the teacher is one of the crucial aspects of using open-ended problems. At times, the teacher may use two class periods for one open-ended problem. In the first period, the students may work individually or in groups to solve the problem and to summarize their findings. Then in the second period, the whole class discusses the approaches and solutions and the teacher gives concluding remarks. From our experience, this teaching approach has proved effective.

IDEAS ON TEACHING THE PROBLEM

When actually teaching the problem, teachers must consider the following points.

Posing the Problem

When open-ended problems are posed in the classroom, students are often asked such questions as "What properties (relations, rules, methods, etc.) can you find?" Such questions may be confusing to some students in the early stages of using this approach because they are not familiar with the use of the terms *property, relation, rule, method*, etc., in mathematics or in responding to such problems and therefore cannot understand what they are expected to do. In helping students understand the meaning of the problem, the following approaches are effective:

1. Encourage students to focus on the same issue by projecting the problem on a screen using an overhead projector and a well-prepared transparency.

2. Add more data for generalization, for example, by introducing variety in the problem situation or by showing more concrete data (examples) than those given in the statement of the problem. For instance, by giving other examples in the problem of the baseball standings, the teacher clarifies relations that are true not only in the given table but also in similar tables.

3. Give examples that do not restrict the students' ways of thinking about the problem. For example, in the problem in which students form, from a given set of solids, sets of solids that have a common property, the teacher may suggest a direction for the work by stating, "At first, let's consider only those that are enclosed by planes."

4. Make good use of such concrete materials as models. The water-flask problem mentioned in chapter 2 is an example.

Organizing the Teaching

Since the open-ended approach places special emphasis on the mathematical thinking of individual students, the teacher must be careful not to impose a particular orientation on all students by adopting the opinions of particular students. This style of teaching, like ordinary teaching, consists of a combination of two things: (a) individual work, and (b) discussion by the whole class. However, since we are not seeking a single solution, we can expect that a new point of view, one that has not yet occurred to students, will emerge at the stage in which the lesson proceeds from individual learning to a class discussion. It is crucial in this approach to proceed from individual learning to group learning.

Recording Students' Responses

It is important to have a written record of the responses, approaches, or solutions to the problem that are taken by each individual and group for later study. Thus, using a notebook or worksheets may be a convenient way for students to record this information. Also, by collecting the worksheets after the lesson, the teacher can use them in evaluating individual and group learning. Since the students' activities at this stage are crucial to further development of the lesson, the teacher should try to identify those students who do not understand the problem and give more examples or suggestions to stimulate them to think in a relevant way about the problem. This can occur while the teacher walks around purposefully scanning students' work. Sufficient time should be allowed for students to complete their work.

Summarizing What Students Have Learned

The teacher or students should write their individual or group work on the chalkboard for all to see. Further, the teacher should include all student propositions even though some may be similar to, or duplicates of, others. Students should be encouraged to confirm whether their work is consistent or can be reduced to a single proposition together with other students. Even when some propositions are erroneous or incomplete in expression, the teacher should regard them in a positive manner and modify them by comments from other students. When the students contribute too many propositions to summarize briefly, the teacher should concentrate on one point of view and lead to a conclusion. Thus, while appropriately incorporating and modifying students' responses, the teacher should integrate and arrange them in order according to particular points of view, summarize the learning, and facilitate a smooth transition to the next lesson.

CRITERIA FOR EVALUATION

Since there will likely be a variety of student reactions or responses to an open-ended problem, it may be difficult for the teacher to evaluate them and make good use of all of them in the lesson. Accordingly, we adopted the following method to evaluate students' activities.

In advance, the teacher prepares a table of expected responses, which are classified and arranged in order, item by item, according to their mathematical features. During the lesson, students' actual responses are checked and entered in a corresponding blank cell of the table. The students' achievement is then evaluated by using this table according to the following criteria:

Fluency—how many solutions can each student produce?

If a student's (or a group's) response is correct from a certain point of view, the teacher awards the student (or group) 1 point. The total of these points is called "the total number of responses." This number can be regarded as an indication of the fluency of students' mathematical thinking.

Flexibility—how many different mathematical ideas are discovered by students?

The correct solutions or approaches produced by one student (or group) can be divided into several categories. If two solutions (or approaches) have the same mathematical idea, they are included in the same category. The number of these categories is called "the number of positive responses." This number can be regarded as an indication of the flexibility of the students' mathematical thinking.

For problems that have several correct answers, we can say that the higher a student's score, the richer is his or her flexibility, or scope, of mathematical thinking.

Originality—to what degree are students' ideas original?

If a student (or group) comes up with a unique or insightful idea, the originality of the idea should be evaluated highly. Among expected responses, several levels of mathematical significance may exist, from higher to lower. The teacher should give a high score to an idea with a high quality of mathematical thinking. The total number of these scores is called "the weighted number of positive responses." This number can be regarded as an indicator of the originality of a student's (or group's) idea.

The first two are methods of evaluating quantity ("how many?"). The third criterion is a method of evaluating quality ("how innovative?"). In our research, we found that students with experience in the open-ended approach received higher scores in flexibility and originality than students with no such experience.

Another criterion for evaluation is the degree of elegance in students' expression of their ideas. Some students write their solutions in ambiguous ways, whereas others do so in a simple, clear, and elegant manner. Elegance in expressing mathematical relations using formulas with words as variables would be better than using ordinary sentences. Using algebraic expressions would be better yet. However, it may be difficult to evaluate objectively the degree of elegance of students' expressions. This criterion will be incorporated into our evaluation system in the future.

Chapter 4

Examples of Teaching in Elementary Schools

I N THIS and the following two chapters, the format described below will be used to report on the experimental lessons we tested in schools. (Japan has a 6-3-3 structure for its schools: six years for elementary school (grades 1–6), three years for lower secondary school (grades 7–9), and three years for upper secondary school (grades 10–12).) Because of space limitations, records of classroom teaching are included only for typical examples. When possible, similar problems are added for the reader's information.

THE FORMAT OF EACH SECTION

The Problem and Its Context

1. *The problem:* The problem is stated in the way it was presented to the students.
2. *Pedagogical context:* The purpose of the problem is given, along with its connections to the content in the textbook and to the mathematics program.

Expected Responses and Discussion of Them

3. *Examples of expected responses:* Examples of students' expected responses to the problem are classified by viewpoints, including some high-quality responses that some students may make.
4. *Discussion of the responses:* The classification of students' responses, the mathematical values of their responses, how to evaluate the responses, and further mathematical development are detailed.

Record of the Classroom Teaching

5. *Teaching the lesson:* The place of the problem in the whole teaching plan, major questions, and related learning activities are discussed. (Minute details are omitted.)
6. *Remarks after the lesson:* Reflection on the lesson, the time needed, classroom discussion, collecting students' responses, and further development of the problem are presented.

INTRODUCTION TO THE IDEA OF PROPORTION

MASAMI TAKASAGO
Elementary School at Yamagata University

The Problem and Its Context

The problem

An insect is walking along a ditch. The chart shows the time required to walk the given distances. The asterisk indicates the distance we forgot to record.

Time (min)	1	2	3	4	5	6	7	8	9	10
Distance (cm)	12	24	36	48	60	72	84	*	*	120

1. What number is represented by the * under 8? Write down the expression you used to find the number.

2. Find another expression you can use to find the number. Write down as many different expressions as possible.

Pedagogical context

The idea of proportion is usually introduced in the following way:

a. The teacher gives the relation of two variables (x, y).

b. The teacher helps students to see that if x is two times, three times, four times (or one-half, one-third, one-fourth) the original value, then y is also two times, three times, four times (or one-half, one-third, one-fourth) the original value.

c. The teacher introduces the phrase "x is in proportion to y" or "x and y are proportional" to refer to such a relation.

However, this approach has two disadvantages:

1. Students may not be interested in, or feel the need to study, such a relation between two variables.

2. From the outset, the teacher goes straight to the second step, which discourages a variety of students' ideas from appearing.

We presented this problem because it allows for a variety of students' ideas and avoids the disadvantages given above. First, we ask the students to find some numbers; then we help them to realize that there is more than one approach to, or way of thinking about, this situation. After examining each approach, students find one common mathematical relationship among the approaches, which helps them to understand the meaning of proportion.

Expected Responses and Discussion of Them

Examples of expected responses

1. Students notice that if the time increases (decreases) by 1 min, then the distance increases (decreases) by 12 cm.

      a. $12 + 12 + \cdots + 12$

      b. $84 + 12, \ 72 + (12 \times 2), \ 60 + (12 \times 3), \cdots$

      c. $120 - (12 \times 2)$

2. Students notice that if the time is two times, three times, . . . the original, then the distance is two times, three times, . . . the original:

      d. $12 \times 8 \quad 24 \div 2 \times 8 \quad 36 \div 3 \times 8 \cdots$

      e. $48 \times 2 \quad 24 \times 4$

      f. $120 \times (8/10)$

3. Students notice that distance (cm) ÷ time (min) is a constant:

      g. $\square \div 8 = 12$

Discussion of the responses

The students' responses can be grouped according to the relations discovered:

1. The amount of increase (decrease)

2. The relative increase (decrease)

3. The relation of two variables

These relations may not be obvious at first. However, when students engage in activities using the chart given in the problems, these relations may be discovered according to the ways in which the students view the chart.

Relation 1 would appear by successive checking:

   a) Starting from the   *b)* Starting from   *c)* Starting from
        beginning   halfway   halfway

Relation 2 appears by checking at intervals:

   d) Starting from the   *e)* Starting from   *f)* Starting from
        beginning   halfway   halfway

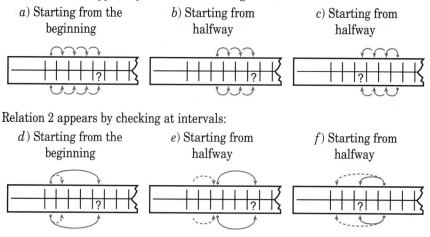

Relation 3 appears by comparing vertically:

g) Comparing the columns

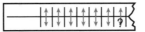

The schemata of the procedures used by students to find the number have no mathematical significance in themselves. However, it is useful to use them to explain ways to find the number, since visual aids facilitate understanding.

When the problem is used in classroom teaching, the expressions offered by the students should be carefully examined. The teacher should make the essential idea included in each expression clear, using the schemata above, and should also help students notice that when they consider the essential ideas of the expressions, the methods of finding the numbers can be classified according to the following three properties:

1. The amount of increase (decrease)

2. The relative increase (decrease)

3. The relation of two variables

The foregoing is an outline of the lesson presented. However, not all aspects of the concept of proportion are clarified in this lesson. Although properties 2 and 3 are characteristics of the proportional relation, at this stage they appear as properties of the proportional relation, not as characteristics to distinguish it from other relations. To make the latter point clear, a further lesson would be necessary to compare the proportional relation with relations having property 1 but not properties 2 or 3. Having the students compare the relations discovered in table 4.1 with those they discover in the chart in the foregoing problem is an example of such a lesson.

Older brother's age	5	6	7	8	9	10	11	12
Younger brother's age	1	2	3	4	5	?	?	8

Table 4.1

Record of the Classroom Teaching

Teaching the lesson

The purpose of this lesson was to introduce the idea of proportion. The concept of ratio was taught before this lesson. In this lesson, the students find relationships between the two variables but the main purpose is not necessarily to make the concept of proportion completely clear. We adopted this approach as an introduction to proportion for the reason mentioned above. The lesson proceeded as follows:

1. The teacher prepared the handouts on which the problem was printed, distributed them to the students, and explained the meaning of the problem. In explaining the problem, the teacher used transparencies on which the table was printed and the numbers in the table were projected on the screen successively one by one. Then the teacher asked the students to devise ways to find the missing numbers.

2. The students presented the expressions they discovered. Those found on the first

attempt are given below (the number in parentheses indicates the number of students in the class of thirty-five students who discovered the expression):

- 12×8 (22)
- $84 + 12$ (11)
- 48×2 (1)
- $12 + 12 + 12 + 12 + 12 + 12 + 12 + 12$ (1)

The expressions that the students found later are these:

- $84 + 12$ (17)
- $120 - (12 + 12)$
- $120 - 12 \times 2$ (17)
- 24×4 (16)
- 12×8 (13)
- 48×2 (13)
- $72 + 12 \times 2$
- $60 + 12 \times 3$
- $48 + 12 \times 4$ (13)
     ⋮
- $120 \times 8/10$ (2)
- $120 \div 10 \times 8$ (2)
- $12 + 12 + 12 + 12 + 12 + 12 + 12 + 12$ (1)

3. The students discussed the ideas represented in their expressions and drew the following conclusions for each of the first four expressions:

- 12×8

a) For every 1 min that passed, the distance increased by 12 cm.

b) When 2 min passed, the distance was 24 cm, which is twice 12 cm. When 3 min passed, the distance was 36 cm, which is three times 12 cm. In the same way, we can multiply by 4, 5, and so on.

c) $24 \div 2 = 12$, $36 \div 3 = 12$, $48 \div 4 = 12$, and so on; so when 8 min has passed, $\square \div 8 = 12$; therefore, the distance is 12×8.

- $84 + 12$

a) When 7 min has passed, the distance is 84 cm; so when 8 min has passed, the distance should increase by 12 cm.

- 48×2

a) When 4 min has passed, the distance is 48 cm; so when 8 min passed, the distance is twice 48 cm.

- $12 + 12 + 12 + 12 + 12 + 12 + 12 + 12$

a) The insect walks 12 cm in 1 min, so we should repeatedly add 12 eight times.

A similar procedure was followed for the second group of expressions, but the details are omitted.

4. The students summarized the underlying ideas by classifying the expressions into groups that were based on the same idea. They found four groups:

a) The group of 12×8: increasing by 12 cm in one minute,

- $12 + 12 + 12 + 12 + 12 + 12 + 12 + 12$
- $120 \div 10 \times 8$

b) The group of $84 + 12$: an increased amount from some starting point

- $72 + 12 \times 2$
- $60 + 12 \times 3$
- $48 + 12 \times 4$

  $\vdots$

c) The group of 48×2: how many times it is from some starting point

- 24×4

d) group of $120 - 12 \times 2$: a decreased amount from some starting point

Thus, most expressions that seemed unrelated at first glance were classified into four groups on the basis of their underlying ideas. (*Note*: In this discussion, $120 \times (8/10)$ was classified into the group of 12×8 because the idea of the expression $120 \times (8/10)$ is similar to that of $120 \div 10 \times 8$.)

5. The students observed the chart given in the problem in order to formulate the rules that were suggested by the information in the chart. The expressions discussed so far were made by using some relations in the chart. These relations were discussed as follows (*T* represents the teacher and *S*, a student):

T: What rule do you notice in this chart?

S: As the time passes, the distance increases by 12 cm.

T: But in the expression $120 - 12 \times 2$, the distance decreases, doesn't it?

S: In the chart, if we look from left to right, the distance increases by 12 cm. If we look from right to left, it decreases.

T: When you look at the chart from left to right, what happens to the time?

S: The time passes.

S: I got it. If the time increases, the distance increases. And if the time decreases, the distance decreases, too.

(The teacher writes rule 1 on the chalkboard.)

> *1. If the time increases (decreases), the distance increases (decreases).*

T: Which expressions are implied by this rule?

S: • 12×8

- 12 + 12 + 12 + 12 + 12 + 12 + 12 + 12
- 84 + 12, 72 + 12 × 2, 60 + 12 × 3, and so on
- 120 – 12 × 2, 120 – (12 + 12)

T: The expression 48 × 2 is not included in this group, correct? Can you see if there are other rules?

S: I think the rule involving multiplication.

T: I am not sure what you mean.

The expression 48 × 2 can be illustrated as in the example below.

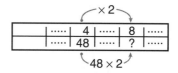

The expression 24 × 4 can be illustrated as in the example below.

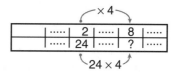

What rule would you propose from these explanations?

S: I got it. If the time is multiplied by 2, then the distance is multiplied by 2. If the time is multiplied by 4, then the distance is multiplied by 4.

S: That's right.

T: Can you say it more simply?

S: If one is multiplied by 2, the other is multiplied by 2.

T: Only in the case of multiplication by 2?

S: By 2, and by 4, too.

S: By 3, too.

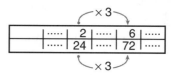

S: Isn't it true when multiplied by any number?

S: Yes, by any number.

T: We can summarize the rules in the following way, correct? (The teacher writes rule 2 on the chalkboard.)

> 2. *If one number is multiplied by 2, 3, 4, ... , then the other is also multiplied by 2, 3, 4, ...*

Note that "…" means "and so on," and that you can continue to put numbers there.

S: I have a question. Since 12×8 is implied by the rule of multiplication by 8, should this expression be in the group with 48×2 and 24×4?

S: I think so.

S: But it was also implied by the rule of increase of 12 cm, too.

(The discussion shows that 12×8 has two meanings.)

T: Can you find another rule?

All the expressions offered by the students were based on either rule 1 or rule 2. The students could not see that the ratio of two variables is a constant. The teacher addressed a student who seemed to have a somewhat different idea.

T: How did you find 12×8?

S: $24 \div 2 = 12$

$36 \div 3 = 12$

$48 \div 4 = 12$

$\vdots$

$\square \div 8 = 12$

So I thought $\square$ is 12×8.

The teacher then helped the students consider the rule derived from this idea. At last, they noticed that the distance divided by the time is always 12. (The teacher writes rule 3 on the chalkboard.)

3. (Distance) ÷ (time) is always 12.

6. As an exercise, the teacher assigned the problem of finding the distance when the time is nine minutes by applying rules 1–3.

Remarks after the lesson

This lesson would have degenerated into a pointless one if it had ended when the students had identified multiple expressions to find the number. But in this lesson the teacher asked the students to classify the expressions into groups based on a common idea and to identify the rules implied in the chart. The lesson was successful because the students keenly tackled the problem, although it took more time.

One reason why the lesson was not completed in one class period was that the teacher allowed students abundant time to find several expressions, and more time was required than expected to discuss these expressions.

A Similar Problem: The Congruence and Similarity of Figures

In the figure shown, *ABCD* is a rectangle. We draw a line *EF* through *O*, the intersection of two diagonals.

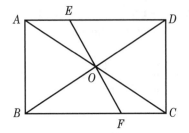

1. Find as many different geometrical figures as possible in this picture. What are the names of these figures?

2. Take two figures among those you found in question 1. Do you observe any rules concerning their sizes or their relative position? Look for other such pairs of figures and the relations among them.

THE CLASSIFICATION OF SOLID FIGURES

KOZO TSUBOTA
Fukazawa Elementary School, Setagaya Ward, Tokyo

The Problem and Its Context

The problem

Several solids are shown in figure 4.1. Choose the solids that share characteristics with solid B and write down the characteristics.

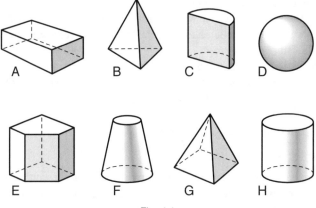

Fig. 4.1

Pedagogical context

When we use the classification of solid figures as an open-ended problem in the classroom, one purpose is to help the students understand that solids can be classified in several ways according to different characteristics. Another purpose is to help them decide on an approach by themselves; that is, to develop flexibility in students' thinking through systematically using all the content they have learned so far and introducing diversity into the problem situation, thereby summarizing the knowledge that they have acquired of solids. Moreover, through the checking of the approaches and classifications that students make, their achievement can be evaluated.

Generally speaking, this topic is treated in textbooks as in the following two examples:

- "Here are several solids [as shown in fig. 4.1]. Which are pyramids? Which are cylinders?" And so on. (Ukita et al.; *Arithmetic 6*, vol. 1, p. 107. Published by Kyoiku Shuppan, 1976.)

• "We classified solids into two sets, A and B. We cut all these solids by a plane that is parallel to the base. In which set is a cross section congruent to the shape of the base?" (Kodaira et al.; *Arithmetic 6*, vol. 1, p. 49. Published by Tokyo Shoseki, 1976.)

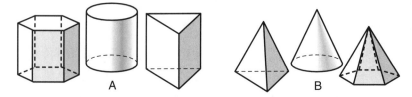

In these problems, the approach is already given to the students and they need not devise their own. The activity is narrowed simply to identifying some figures that have a certain property. If we modify these textbook problems and make them into open-ended problems, we can naturally stimulate students' flexibility in thinking.

Although we have planned this problem to be a summary lesson for learning solids, the same problem can also be used as an introduction for learning about solids. It is possible to develop a lesson plan by noting the most important student responses and making good use of them in planning the lesson. Students would be pleased to know that their reactions are incorporated into the lesson, and their willingness to study would be encouraged. If we use this problem as both an introduction and a summary, we can evaluate each student's learning through the lesson.

In this problem, we have specified that the figures be compared with figure B. Another approach is to have students choose one figure they like and proceed in a similar way. (See the "Similar Problem" that follows at the end of this section.)

Expected Responses and Discussion of Them

Examples of expected responses

Students' responses are expected to vary a great deal. Rarely would they be arranged in order according to their mathematical quality. Table 4.2 indicates which characteristics we can expect students to cite for solid B at the beginning of this chapter. A slash means that the solid has the characteristic.

Characteristics / Solids	A	(B)	C	D	E	F	G	H
Pyramid		/					/	
Having triangular faces		/					/	
Having four total faces (sides and base)		/	/					
The view from the side being a triangle		/					/	
Faces are all polygons	/	/			/		/	
Cross section parallel to the base being similar to, but not congruent to, the base		/				/	/	

Table 4.2
Expected Responses with Respect to Solid *B*

At first, we think about the solid B. However, if solid H is chosen, students may find more of a variety of viewpoints than with B. It would be more effective, then, to treat solid B as a "warm-up" and to treat solid H as a main topic to scrutinize more deeply. The expected student responses for solid H are shown in table 4.3.

Characteristics / Solids	A	B	C	D	E	F	G	(H)
Solid of revolution				/		/		/
Cylindrical solid	/		/	/				/
When viewed from the top, the shape being a circle				/		/		/
Side face being a rectangle	/		/	/				/
Having three faces						/		/
Cross sections parallel to the base being congruent	/		/	/				/
Side surface being curved						/		/
Front view being a rectangle	/		/	/				/

Table 4.3
Expected Responses with Respect to Solid H

Discussion of the responses

We consider students' responses from two perspectives—quantitative and qualitative. Quantitatively, the responses are evaluated by using the number of positive responses, which means that the features referred to are mathematically correct. Qualitatively, the responses are evaluated by using a classification table that is prepared before the lesson (e. g., table 4.4). In the table, expected student-responses are classified into several groups that were organized in advance according to their mathematical characteristics. Students' responses are evaluated by the number of groups that include their responses. Table 4.4 is one example of such a classification.

Classification Number	Contents
1	Shape of faces (side and base)
2	Number of edges, vertices, faces, and relations among them
3	Parallel or perpendicular relations between edges and faces
4	Shape of a projection (plane, front view, side view)
5	Shape of a cross section (by a plane parallel, perpendicular, or slanted to the base)
6	Shape of a development of the solid
7	Construction by moving a plane-figure (rotation or parallel movement)
8	Volume, surface area
9	Others (angles between side and base planes, curved or flat, cylindrical or conical)

Table 4.4
Classification Table

Because there are two aspects to consider (qualitative and quantitative), it is desirable that the students give many responses for both aspects. Furthermore, if the teacher gives the same problem at the beginning and end of a certain time-interval, the teacher can evaluate changes in each student's achievement by considering changes in the number of responses mentioned above and the improvement or refinement in writing their responses.

Record of the Classroom Teaching

Teaching the lesson

This problem was posed to sixth graders in the fourteenth and final lesson of a unit on solid figures as a summary of what students had learned.

The following chart presents the progression of the lesson.

Learning Activities	Questions to the Students and Expected Responses	Remarks for the Teacher
Understanding the meaning of the problem	Determine the solids that share the same character-istics with solid B and write them down. (See fig. 4.1.)	Distribute prepared handouts.
Finding the solids that have the same characteristics as solid B.		Display a poster board on which the solids are drawn.
		Students write on their own handouts.
Presenting students' responses and discussing them	Let's make a table of your responses.	Fill in the table according to students' responses.
Classifying the solids by considering several viewpoints	Choose solid H and solve the same problem.	Distribute another hand-out and have students fill it in. (A blank table is printed on the handout.)
Presenting students' responses and discussing them	Share your responses and let's make a table of them.	Fill in the table according to students' responses.
		Let students begin with characteristics not shared with solid B.
Summarizing	By a change of viewpoint, a solid can be regarded as belonging to the same group as a different solid.	
	In this lesson, we have reviewed all the topics we have learned so far.	

This lesson was videotaped so that a record (see table 4.5) of the elapsed time for each part could be kept.

Time Elapsed (min)	Time	Activities
3	3	Teacher distributes handouts. A student reads the problem aloud. Teacher explains the problem.
7	10	Students work on the problem.
18	28	Students present their responses for discussion.
2	30	Teacher distributes the other handouts.
7	37	Students work on the problem.
15	52	Students present their responses for discussion.
2	54	Teacher summarizes the lesson.

Table 4.5
Record of Time in the Lesson

Students' responses for solid B and solid H are given in tables 4.6 and 4.7, respectively.

Characteristics / Solids	A	(B)	C	D	E	F	G	H
Pyramid		/					/	
Having only one base		/					/	
Side being a triangle		/					/	
Surface being flat	/	/			/		/	
Cross section parallel to the base being similar to the base		/				/	/	
Not a solid of revolution	/	/	/		/		/	
Having no face parallel to the base		/					/	
Cross section perpendicular to the base through the vertex being a triangle		/					/	
Edges having straight lines only	/	/					/	
Having volume	/	/	/	/	/	/	/	/
Shape of shadow being a triangle		/					/	
Having Vertices	/	/			/		/	
Having four faces		/	/					
Having a perpendicular segment from the vertex to the base as its height		/					/	
Number of edges = (Number of edges of the base) × 2		/					/	
Viewed shape from the top being a polygon	/	/			/		/	

Table 4.6
Students' Responses with Respect to Solid B

Characteristics / Solids	A	B	C	D	E	F	G	(H)
Having one side face						✓		✓
Having two bases	✓		✓		✓	✓		✓
Base being perpendicular to side face	✓		✓		✓			✓
Having no vertex				✓		✓		✓
Viewed face from the top being a circle				✓		✓		✓
Being a solid of revolution				✓		✓		✓
Cross section parallel to the base being a circle				✓		✓		✓
Cross section not parallel to the base being an ellipse						✓		✓
Being a cylindrical solid	✓		✓		✓			✓
Having parallel side faces	✓		✓	✓	✓			✓
Side views being a quadrilateral	✓		✓		✓			✓
Having three faces						✓		✓
Having two edges						✓		✓
Having side faces	✓	✓	✓		✓	✓	✓	✓
Cross sections parallel to the base being congruent	✓		✓		✓			✓
Cross section perpendicular to the base being a rectangle	✓		✓		✓			✓
Viewed shape from the top being a circle				✓		✓		✓
Enclosed by faces	✓	✓	✓	✓	✓	✓	✓	✓

Table 4.7
Students' Responses with Respect to Solid H

In table 4.8, students' responses were classified into nine groups according to the criteria given in table 4.4. (The number of students was 38.)

Classification number	1	2	3	4	5	6	7	8	9
Number of students responding	7	36	7	20	18	0	16	1	27
Number of positive responses	7	77	9	28	26	0	17	1	37

Table 4.8
Students' Responses to the Problem of Solid H

Remarks after the lesson

Originally, time was allocated for each problem as follows:

- First problem (10 min), discussion (15 min)

- Second problem (10 min), discussion and summary (10 min)

However, the lesson required fifty-four minutes because students actively participated in discussing the two problems.

The teacher prepared a large poster on which the students' responses were written. Many students, including some who were usually inactive, wanted to present their ideas. In this lesson, more than half the students spoke. For solid B, they were so keen on presenting their ideas that the teacher had to end this part of the lesson by saying, "We will stop the discussion after just three more presentations." Their enthusiasm was due to the teacher's respect for, and acceptance of, all responses, which is a characteristic of this approach to teaching. In ordinary lessons, teachers are likely to disregard students' less desirable ideas, even though they may not be incorrect. For this reason, the teacher had to increase the time allocated for solid H. Observing the videotape confirmed that below-average students significantly contributed to the lesson.

Figures 4.2 and 4.3 show the student responses to this problem. Each circled number on the graphs indicates the number of students who responded; the x-coordinate indicates the number of categories and the y-coordinate the total number of positive responses. In both figures, the larger the values of the coordinates, the better the results. (Results give, of course, students' achievement only relative to their peers.) The same problem was given to the same students again four months after the lesson. From the two graphs, it is clear that a right-upward movement is indicated, though there were a few students who showed no improvement when scrutinized individually.

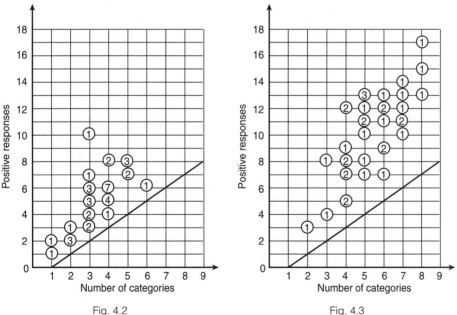

Fig. 4.2
Students' responses at the end of the lesson

Fig. 4.3
Students' responses four months after this lesson

It would have been more desirable to group the students' responses by as many viewpoints as possible in the summary stage. Further, we should have focused on either one of the two problems, since time was short toward the end of the lesson.

A Similar Problem: The Classification of Plane Figures

Look at the figures in figure 4.4. Find as many different characteristics as possible that create a group of figures that share the same characteristic. Classify these figures according to those characteristics.

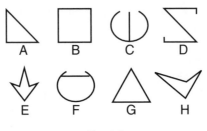

Fig. 4.4

For example, if the characteristic "having right angles" is noticed, then figures A, B, and D share this characteristic. Accordingly, you should make slash marks as indicated in the table below.

Characteristics / Figures	A	B	C	D	E	F	G	H
Having angles of 90°	/	/		/				

RANKING TEAMS IN A MARATHON RACE

KANJIRO KOBAYASHI
Elementary School at Chiba University

The Problem and Its Context

The problem

Three teams, A, B, and C, participated in a marathon race. Each team had ten runners. The results are given in the chart below. Which team do you think was the winner? Find as many ways as you can to determine the winner.

Runner's Ranking	1	2	3	4	5	6	7	8	9	10	11	12	13	14	15
Runner's Team	A	B	A	C	B	B	C	A	C	C	C	B	A	A	B
Runner's Ranking	16	17	18	19	20	21	22	23	24	25	26	27	28	29	30
Runner's Team	B	C	A	C	B	C	B	B	A	C	A	A	A	C	B

Pedagogical context

It is difficult to agree on how to rank teams in competitions on the basis of individual results. Several different methods can be used to decide on the winning team. In this problem, students are expected to find as many methods as possible.

Since this type of problem is not explicitly included in the official Japanese course of study put in place in 1968, it is not included in typical textbooks. However, because the essential part of this problem is considered a measure of the tendency of a group, which is described in the course of study, the problem could be included in the fifth grade under the topic of statistics.

Expected Responses and Discussion of Them

Examples of expected responses

Rankings can be made in many ways. We divided them into two categories.

- *Ranking by performance of a subset of members of each team. Nine ways to do so follow:*

1. Ranking by the number of runners each team had in the top ten places

Team	No. of Runners	Team's Ranking
A	3	Second place
B	3	Second place
C	4	First place

2. Ranking by the order of total scores of each team in the top ten, which are obtained by assigning 1 point to the first runner, 2 points to the second runner, and so on up to the tenth runner (lowest total is best)

Team	Runners' Scores	Team's Ranking
A	$1 + 3 + 8 = 12$	First place
B	$2 + 5 + 6 = 13$	Second place
C	$4 + 7 + 9 + 10 = 30$	Third place

3. Ranking by the average scores for the three groups in method 2 above (the result is the same)

4. Ranking by the top runner in each group

The best runner in Team A was first	First place
The best runner in Team B was second	Second place
The best runner in Team C was fourth	Third place

5. Ranking by the last runner in each group (lowest is best)

The last runner in Team A was 28th	First place
The last runner in Team B was 30th	Third place
The last runner in Team C was 29th	Second place

6. Ranking by the total scores or average scores of the top five runners in each team, by assigning 1 point to the first runner, 2 points to the second runner, and so on up to the fifth runner

Team A: $1 + 3 + 8 + 13 + 14 = 39$	First place
Team B: $2 + 5 + 6 + 12 + 15 = 40$	Second place
Team C: $4 + 7 + 9 + 10 + 11 = 41$	Third place

7. Ranking by the median ranks (the fifth runner) in each group

The fifth runner in Team A was 14th	Second place
The fifth runner in Team B was 15th	Third place
The fifth runner in Team C was 11th	First place

8. Ranking by the modal rank of the three teams, that is, by grouping ranks 1st through 10th, 11th through 20th, and 21st through 30th

Team \ Class	1st–10th	11th–20th	21st–30th	
A	3	3	(4)	Team A is third
B	3	(4)	3	Team B is second
C	(4)	3	3	Team C is first

9. Ranking by the order of the differences between the first and the last runners in each team (lowest is best)

Team A	$28 - 1 = 27$	Second place
Team B	$30 - 2 = 28$	Third place
Team C	$29 - 4 = 25$	First place

- *Ranking by the performance of a whole group. Two ways to do so follow:*

10. Ranking by the reverse order of the total scores of each team, which are obtained by assigning 1 point to the first runner, 2 points to the second runner, and so on up to the tenth runner

Team A	$1 + 3 + 8 + 13 + 14 + \cdots + 28 = 162$	Third place
Team B	$2 + 5 + 6 + 12 + 15 + \cdots + 30 = 151$	First place
Team C	$4 + 7 + 9 + 10 + 11 + \cdots + 29 = 152$	Second place

11. Ranking by the reverse order of the sums of the differences between the rank of each team member and the average rank in the team

Team A (average 16) $15 + 13 + \cdots + 11 + 12 = 84$	Third place
Team B (average 15) $13 + 10 + \cdots + 8 + 15 = 71$	Second place
Team C (average 15) $11 + 8 + \cdots + 10 + 14 = 70$	First place

Other rankings are possible for the three teams; for example, considering the lowest ten runners or the strict mean deviation.

Discussion of the responses

It is very difficult to decide which is the best method because each has its merits and demerits. It is reasonable to include the performance of all participants in the result, since the exercise was a group marathon. It is also reasonable to consider the performance of a subset as representative of the whole group. This is similar to situations in statistics in which there are many ways to determine a central value of a population (e.g., a weighted mean or a median) depending on the purpose or use to be made of it. The mathematical values of the methods will be discussed below and some remarks on the lesson will be given.

Ranking by the performance of a subset of a group. The most likely student response is number 1 above; that is, the number of runners in the top ten places. A team may be ranked high if it includes several fast runners, whereas others may be slower. Therefore, this response cannot reflect the positions of all runners. It is useful, however, when our interest is concerned with which team has more good runners.

For method 1 above, a team monopolizing the top three places will be ranked lower than a team having four runners in the seventh to tenth places. To remedy this, methods 2 (by the total sum) or 3 (by the average) may be plausible.

Method 4 ranks performance by only one runner, that is, the top one in each team, and others are ignored. Method 5 ranks performance by the slowest runner, which is essentially the same as 4.

Method 6 ranks the teams according to the total score of the top five runners in each team. This reflects the performance of each team better than methods 1 or 2, but the performances of runners in the lower positions are ignored. Method 6 is similar to the scoring system used in group competition (e.g., gymnastics) in the Olympic Games.

Method 7 is based on the medians of each team. Therefore, it involves the performance of both a subset of each team and the entire team. This may seem appropriate at first, but this method does not reflect the deviation within a group. Method 8 improves on method 7; it is based on modal rank and involves the peak performance distributed among the three teams. However, this may be difficult for elementary students to observe.

Method 9 ranks the teams according to the extent of clustering in each team as a whole. However, in the instance of an extreme case, the method may be weakened.

Ranking by the performance of all team members. The most likely student response is method 10 above. It incorporates the performance of each member in a group. Where the numbers of runners in the teams are unequal, a weighted total can be used as an alternative.

A larger total score does not necessarily imply greater clustering. Method 11 attends to this aspect and permits a comparison of the tendencies, within each team, among the three teams. However, it does not yield a sophisticated measure of dispersion.

The objective of the lesson is not only to encourage students to discover several or many approaches to solving the problem but also to address the merits and demerits of each. Not all methods enumerated above are expected to occur in a lesson. The approaches that the students generate may depend on the grade level or the educational level of the students. But the teacher should anticipate as many approaches as possible as part of the lesson plan when using the open-ended approach.

This problem can be extended by considering unequal groups and having students then consider how the rankings can be made.

Record of the Classroom Teaching

Teaching the lesson

Because this problem is not directly connected with the official Japanese course of study, it can be taught at any time as an independent topic. Since elementary descriptive statistics is taught to fifth-grade students in Japan, it seems appropriate to teach this lesson as an application after the teaching of statistics is completed. However, since students' learning and experience are not deep at this level, special consideration should be given to such challenging problems, especially how to promote students' interest and understanding and how to deal with various responses of students from a pedagogical point of view.

The lesson was taught at the end of the last semester of the fifth grade. The following points were emphasized:

- The teacher encouraged active discussion within and among groups of students in order to discover as many methods of solution as possible.
- The teacher encouraged discussion of the merits and demerits of each method in order to clarify the mathematical features of each.
- The teacher encouraged students to focus on the common properties among the methods, which enabled students to classify the methods into categories.
- The teacher used rhetorical questions that ostensibly narrowed students' range of thinking when widening was desired or, alternatively, widened their thinking when narrowing was desired (e.g., "There is this method only, right?" or "All these methods are different from each other, right?").

The lesson proceeded as shown in the chart in figure 4.4.

Teacher's Questions	Learning Activities
• What games do you play outside with your classes?	• Talk about baseball, swimming, races, and so on.
• How do you decide which class finishes first?	• Help students realize that in individual competitions, such as a swimming meet, it may be difficult to decide which class finishes first.

• Three teams of students competed in a marathon, with the results shown. Which team do you think finished first?	• After reading the problem, students guessed which team was first:
	  * I think team A was the best because two of its students finished first and third.
	  * I think B was first.
	  * How about finding the average?

1	2	3	4	5	6	7	8	9	10	11	12	13	14	15
A	B	A	C	B	B	C	A	C	C	C	B	A	A	B

16	17	18	19	20	21	22	23	24	25	26	27	28	29	30
B	C	A	C	B	C	B	B	A	C	A	A	A	C	B

Teacher's Questions	Learning Activities

• How do you decide the ranking of the teams?	• Thinking individually, students express their ideas by using numbers:
	  * I want to add the rankings.
	  * I will try to find the average.

• Let's make three groups of our class members corresponding to A, B, and C. Can you find methods that might make your group finish first?	• Solving the problem in groups, students devise their methods by discussion and then prepare a report of their ideas.
• Can we determine the merits and demerits of each method?	• If they consider the number of students in the top ten places, then group C is the best, though some members of C cluster in lower places (7th, 9th, 10th). • They next consider clusters.

• Can we have each group present its methods?
 	
(Present the second and third methods.)
	• Students present their ideas to the whole class:
	  * The total of the rankings
	  * The number of students in the top ten places
	  * The ranking of students in the middle

• What are the merits and demerits of each method?	• They find that B finishes first using the average but C finishes first according to clustering.

Fig. 4.4

Teacher's Questions	Learning Activities
What mathematical features did you discover?	

• How can we summarize the lesson? (The teacher gives a summary.)	• They classify the ranking methods into two groups:
	1. Ranking by performance of a subset of each group
	  • The best ranking
	  • The number of runners in the top ten
	2. Ranking by the performance of a whole group
	
	  • The total ranking
	  • The average ranking

• How was today's lesson different from your usual ones?	• There were several different answers and a discussion of the quality of them.

Fig. 4.4 (continued)

There were forty students in all. The number of students who discovered each of the various methods is shown below in parentheses. Some methods emerged from group discussions.

1. Method 1   (22)
2. Ranking the total scores of each group, obtained by assigning 10 points to the first runner, 9 to the second, and so on.   (1)
3. Determining the average score, as mentioned above   (0)
4. Method 4   (3)
5. Method 5   (2)
6. Ranking by the average rank of the top five runners in each group   (3)
7. Method 7   (0)
8. Method 8   (1)
9. Method 9   (14)
10. Method 10   (32)
    • Ranking by total scores for each group, obtained by assigning 60 points to the first runner, 59 points to the second runner, and so on   (11)
    • Ranking by the average for each group   (10)
11. Method 11   (4)
12. Ranking according to the group totals, obtained by assigning 3 points to the first through tenth runners, 2 points to the eleventh through twentieth runners, and 1 point to the rest   (2)
13. Ranking according to the fifth (or the tenth) runner in each group   (3)

14. Ranking according to each three runners having the same rank within each group, as shown in the following table. (This student did not explain anything beyond this. Perhaps the student would decide the group order by the number of ones in the lines of the second table.) (1)

Ranking within a group / Group	1	2	3	4	5	6	7	8	9	10
A	1	3	8	13	14	18	24	26	27	28
B	2	5	6	12	15	16	20	22	23	30
C	4	7	9	10	11	17	19	21	25	29

↓

Ranking within a group / Group	1	2	3	4	5	6	7	8	9	10
A	1	1	2	3	2	3	3	3	3	1
B	2	2	1	2	3	1	2	2	1	3
C	3	3	3	1	1	2	1	1	2	2

15. Ranking by decreasing order of a number obtained by subtracting the number of runners in twentieth through thirtieth place from the number of runners in first through tenth place (sometimes subtraction may become impossible) (1)

16. Ranking by increasing order of the sum of all the differences from the average in each group (1)

Clearly students' thinking about how to solve this problem was diverse. As expected, the most frequent methods used by students were methods 10, 1, and 9, in that order.

It was a pleasant surprise to see that some students noticed and used the idea of dispersion. However, the students' verbal expressions of their thinking were not always simple or easy to understand, as seen in the examples. This is not uncommon in this approach to teaching. Overall, students exhibited more variety in their methods than expected.

Discussion of the responses

If we had merely presented the problem without further facilitation by the teacher, could so many different responses have been expected? We think not. The results would have been limited to at most methods 10 and 1. In the first stage of the lesson, students were not able to discover many methods, and they wrote only one or two in their notes. However, by the end of the lesson, the number of students' methods increased to three or four, largely because of four reasons:

1. Students actively exchanged their ideas and discussed them in their groups.
2. The teacher presented a situation in which students supposed they were in one of the three groups (A, B, C). Then they tried to determine methods through a process of friendly competition.
3. The students found the problem interesting, and they enjoyed the learning.
4. Enough time was given for students to think about the problem.

Moreover, comparing and discussing the methods of quantification was effective in deepening students' understanding of the meaning of average.

The lesson can be improved in a number of ways. It required sixty minutes, which was more than expected. It would be better to use one period just for discovering different methods and a second for focusing on discussion. Furthermore, we could prepare the problem and approach in such a way that students' methods would emerge more quickly, that is, by changing students' fixed idea that finding a total score is the best approach. We need to develop a good questioning approach to stimulate students' natural ways of thinking that lead to multiple approaches to solving the problem.

A Similar Problem: Comparing Weights

There are five objects P, Q, R, S, and T with P the heaviest. The order of the weights is P, Q, R, S, T. Four students, A, B, C, and D, tried to guess the order of the weights as follows:

A:	R	P	S	T	Q
B:	Q	R	P	S	T
C:	T	R	Q	S	P
D:	S	Q	R	P	T

Find as many methods as you can to decide who is the best guesser. According to your methods, who is the best guesser? The second best? The third best?

THE CLASSIFICATION OF GRAPHS

MASAMI TAKASAGO
Elementary School at Yamagata University

The Problem and Its Context

The problem

The following graphs represent the relation between two variables:

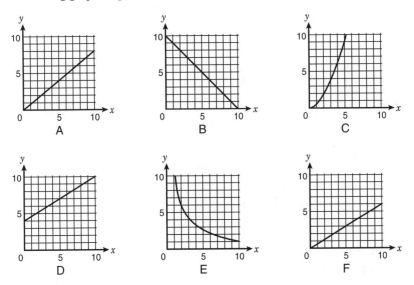

1. Classify these graphs into two groups, one group being a set of graphs that have some mathematical features in common with graph A, and the other that does not. Find as many different methods to classify as you can, and explain the basis for your classification.

2. Choose a different graph and classify the graphs in a similar manner.

Pedagogical context

We use this problem after students have studied direct and inverse proportion and ways to express graphs in the sixth grade. The problem can be used as a summary of students' learning of relations such as proportion, inverse proportion, and the constant sum and constant difference of two variables.

Since students have previously learned features of graphs, solving this problem provides a review and deepens their understanding of them through classification in various ways. Graph C is not studied at the elementary school level. However, we include a table of numbers with this relation reflected in it.

This lesson flows as follows:

1. Classify graphs B through F into two groups: those having a feature in common with graph A and those that do not.

2. Explain the basis for the classification (e.g., visual features, such as the graph rises to the right or is a straight line).

3. Study the mathematical meanings in the methods of thinking about the problem (i.e., the meaning in terms of numerical changes of variables).

4. Study another classification method in the same way.

5. Make a summary of the methods of the classification as shown in the following diagram.

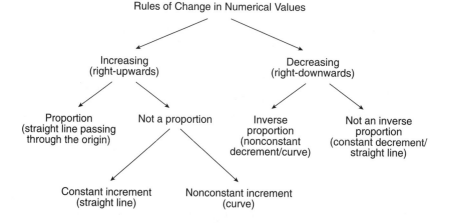

Expected Responses and Discussion of Them

Examples of expected responses

1. Determine a set of graphs based on graph A:

A and B, D, F	(The graphs are straight lines.)
A and F	(The graphs are straight lines through the origin.)
A and C, D, F	(The graphs rise to the right.)
A and C, F	(The graphs pass through the origin.)
A and D, F	(The graphs are straight lines and rise to the right.)
A and C, D, E, F	(The graphs extend indefinitely in the first quadrant.)

2. Other classifications:

C and E	(The graphs are curved lines.)
B and E	(The graphs fall to the right.)
B and D and E	(The graphs do not pass through the origin.)
D and F	(The graphs have the same slope.)

Discussion of the responses

Notice that all the methods above determine a set based on visual similarities between the graphs. Indeed, in this lesson, the teacher encouraged students to think about the common relationships between the two variables through their visual similarity.

Now let us consider the mathematical meanings of the observed relationships.

1. *A group of straight lines and a group of curved lines.* The response that may occur immediately is that one set is formed of graphs that are straight lines. The question here is concerned with what common ideas in the variation of two variables are implied by the visual features. When we make a numerical table representing the relation between two variables (x, y) to study this problem, we know that a graph is a straight line only if the change of y corresponding to the change of x by 1 is a constant, as with graphs A, B, D, and F below:

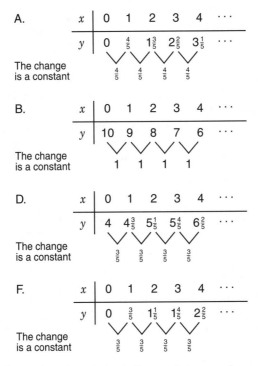

A.

x	0	1	2	3	4	$\cdots$
y	0	$\frac{4}{5}$	$1\frac{3}{5}$	$2\frac{2}{5}$	$3\frac{1}{5}$	$\cdots$

The change is a constant $\frac{4}{5}$ $\frac{4}{5}$ $\frac{4}{5}$ $\frac{4}{5}$

B.

x	0	1	2	3	4	$\cdots$
y	10	9	8	7	6	$\cdots$

The change is a constant   1   1   1   1

D.

x	0	1	2	3	4	$\cdots$
y	4	$4\frac{3}{5}$	$5\frac{1}{5}$	$5\frac{4}{5}$	$6\frac{2}{5}$	$\cdots$

The change is a constant $\frac{3}{5}$ $\frac{3}{5}$ $\frac{3}{5}$ $\frac{3}{5}$

F.

x	0	1	2	3	4	$\cdots$
y	0	$\frac{3}{5}$	$1\frac{1}{5}$	$1\frac{4}{5}$	$2\frac{2}{5}$	$\cdots$

The change is a constant $\frac{3}{5}$ $\frac{3}{5}$ $\frac{3}{5}$ $\frac{3}{5}$

However, if the change is not a constant, the graph is curved, as with C and E below:

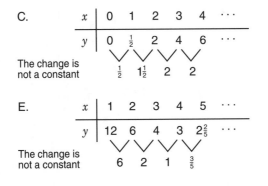

C.

x	0	1	2	3	4	$\cdots$
y	0	$\frac{1}{2}$	2	4	6	$\cdots$

The change is not a constant $\frac{1}{2}$ $1\frac{1}{2}$ 2   2

E.

x	1	2	3	4	5	$\cdots$
y	12	6	4	3	$2\frac{2}{5}$	$\cdots$

The change is not a constant   6   2   1 $\frac{3}{5}$

2. *Graphs that rise to the right and fall to the right.* When we study the meaning of the relations in the tables, we know that—

- if a graph rises to the right, the *x* and *y* values are increasing;
- if a graph falls to the right, the *y* values are decreasing.

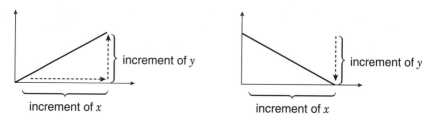

3. *The graph that extends indefinitely and the graph that is limited by the axes.* Previously, the students learned that the proportional relation is graphically expressed by a straight line through the origin. It is important to encourage students to pay attention to the idea that a straight line extends indefinitely, otherwise specified by its range. This lesson provides an opportunity to reconfirm this notion by comparing it with graph B, whose range is restricted to the first quadrant because negative numbers are not in the students' curriculum:

- Graphs extending indefinitely—limitless range
- Limited graphs—finite range

4. *Graphs that have the same slope.* It is clear from the tables for graphs D and F in number 1 that the increments in *y* corresponding to the increments by 1 in *x* are the same.

A Similar Problem: Many Kinds of Graphs

The purpose of making graphs is to represent mathematical relationships visually. The eight graphs that follow express such relationships. Determine as many purposes as possible that are common to more than two graphs.

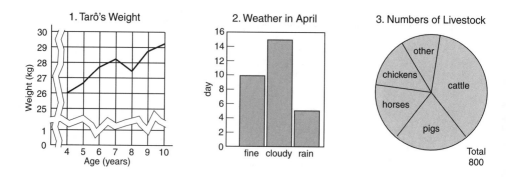

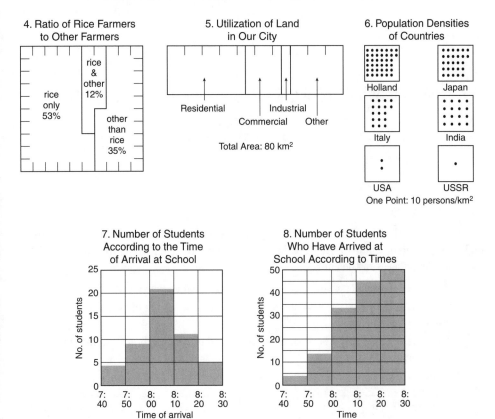

4. Ratio of Rice Farmers to Other Farmers

rice & other 12%
rice only 53%
other than rice 35%

5. Utilization of Land in Our City

Residential | Industrial
Commercial Other

Total Area: 80 km²

6. Population Densities of Countries

Holland   Japan
Italy   India
USA   USSR

One Point: 10 persons/km²

7. Number of Students According to the Time of Arrival at School

No. of students

Time of arrival

8. Number of Students Who Have Arrived at School According to Times

No. of students

Time

REVIEW OF THE LEARNING OF PROPORTION

SHIZUE MORI
Asahi Elementary School, Fukui City

The Problem and Its Context

The problems

   Problem 1. Hiroshi was thinking about the four situations below. His brother, a junior high school student, came in. He filled in a few spaces in the chart in figure 4.5. Try to fill in the remaining spaces in the chart.

   1. I bought a flower for x yen. I paid 100 yen. The change was y yen.

   2. There is a flower bed whose shape is a square. The length of each side is x m. The area is y m^2.

   3. There is a flower bed whose shape is rectangular. Its length is x m, and its width is y m. The area is 36 m^2.

   4. I bought x kg of sugar. The price was 200 yen per kg. I paid y yen.

Situation	1. I bought a flower for x yen. I paid 100 yen. The change was y yen.	2. There is a flower bed whose shape is a square. The length of each side is x m. The area is y m².	3. There is a flower bed whose shape is rectangular. Its length is x m and its width is y m. The area is 36 m².	4. I bought x kg of sugar. The price was 200 yen per kg. I paid y yen.
Table			x: 1, 2, 3, 4, ⋯ y: 36, 18, 12, 9, ⋯	
Algebraic Expression				$y = 200 \times x$
Graph		(graph with y-axis marked 4, x-axis marked 0 1 2, curve)		
Figure Diagram			36 { ... 1	400 / 200 / 1kg / 2kg with ? marks

Fig. 4.5

Problem 2. When Hiroshi finished filling in the spaces in the chart, his sister, a high-school student, came in. She said, "Situations 1 and 3 look alike." How are they alike? Explain why.

Compare the other situations and their relations with one another and find the points that are alike. Explain.

What did you discover by
yourself?

What did your group
discover?

What was discovered in the
discussion by the whole class?

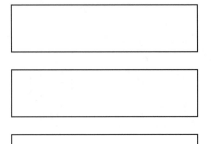

Pedagogical context

These problems are used as a summary of students' learning of relations, where the sum, difference, product, or quotient of two variables is constant. Students previously learned how to produce tables, algebraic expressions, and graphs that express these relations. The purpose of this lesson is to help students understand the importance of observing the relations from different perspectives and to deepen their understanding by studying how the relations are alike or different.

Expected Responses and Discussion of Them

Examples of expected responses

Problem 1. See the chart in figure 4.6.

Problem 2. Relations that are alike among the following situations:

1 and 3:
- When x increases, y decreases.
- The graphs fall to the right and do not pass through the origin.
- $x + y$ is a constant in number 1 and $x \times y$ is a constant in number 3.

1 and 4:
- The graphs are straight lines.
- The rate of increase (or decrease) is a constant.
- Both situations are about money.
- Segment diagrams can be made.

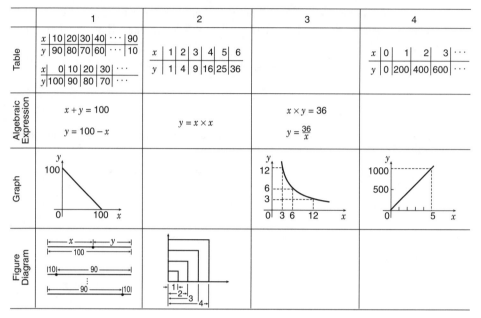

	1	2	3	4
Table	x | 10|20|30|40| ⋯ |90 y | 90|80|70|60| ⋯ |10 x| 0| 10| 20| 30| ⋯ y|100| 90| 80| 70| ⋯	x | 1 | 2 | 3 | 4 | 5 | 6 y | 1 | 4 | 9 |16|25|36		x |0| 1| 2| 3| ⋯ y |0 |200|400|600| ⋯
Algebraic Expression	$x + y = 100$ $y = 100 - x$	$y = x \times x$	$x \times y = 36$ $y = \frac{36}{x}$	
Graph				
Figure Diagram				

Fig. 4.6

2 and 3:	• The graphs are curved lines.
	• Both situations are concerned with area.
2 and 4:	• The graphs go through the origin.
	• The graphs rise to the right.
	• If one variable increases, the other also increases.
	• The range of x is any positive number.
1, 3, and 4:	• For every x there is one y.
1, 2, 3, and 4:	• All the relations are concerned with two variables.

Discussion of the responses

Students are expected to have little difficulty understanding the problems, since they have previously learned proportion. Many students may write the algebraic expression $y = 2 \times x$ in problem 2 instead of $y = x \times x$ (x times x). Although it will be easy for the students to observe the increase or decrease from tables and graphs, it may be more difficult for them to understand why this is so. Students can easily find differences between pairs of relations, but it is important to summarize them as "the relation of two variables."

A Similar Problem: The problem in "The Classification of Solids" section earlier in this chapter

MAKING A GROUP OF RELATED FIGURES

TAKEHISA ASAKURA

Kamegawa Elementary School, Beppu City

The Problem and Its Context

The problem

A triangle is shown below. Draw as many figures as possible that have some common properties with the given triangle. Briefly explain why the properties are common to the given one and the ones you draw.

Pedagogical context

In this problem students are asked to draw figures that have a property in common with the given one. The purpose is to encourage students to recognize mathematical features from different perspectives.

The problem may be used as a problem of classification. However, since the problem is geometric, it is better to use the problem as a summary of learning about "scale drawings of geometric figures" in the fifth grade while also reviewing classification. Some of the students' learning may be further developed to the point where figures and their transformed figures are regarded as the same or equivalent when the latter can be transformed back to the former; for example, with congruent, similar, or topological transformations.

This lesson is specifically intended to encourage students to notice features of triangles other than right triangles as a starting point. Then they learn, for example, that a triangle cannot be both an equilateral triangle and a right triangle but that it is possible for a triangle to be at the same time both a right triangle and an isosceles triangle; that is, a broader concept of equilateral triangle. Furthermore, from a more general point of view, both a square and a circle can be regarded as the same kind of figure in the sense that both are simple closed curves in a plane. These examples may help to enrich students' views of geometric figures.

In a traditional teaching approach to review, all the figures have been prepared and organized beforehand and the classifying criteria given by the teacher. Consequently, the approach is imposing, and students will likely respond passively. If we change the problem to an open-ended one, however, then the students will respond in an active way, and the properties of triangles will be learned spontaneously from their responses and the discussion. Furthermore, this approach provides an opportunity for students to learn more than what is prescribed at this grade level and has the further merit of inducing a variety of mathematical ideas from the students.

Expected Responses and Discussion of Them

Examples of expected responses

Students are expected to draw figures like those in figure 4.7.

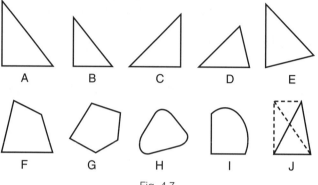

Fig. 4.7

Students will adopt the following viewpoints from which common properties are identified:

Viewpoint relating to—		Common Properties
Shape	A:	The two figures (A and the given figure) are congruent.
	B:	Figures are similar.
	C:	The triangles have a right angle.
	D:	The figures are triangles (any shape).
Side	A:	The triangles have corresponding sides equal (congruence).
	B:	The ratio of corresponding sides for the two triangles is constant (similarity).
	D:	The figures are bounded by three straight lines (triangle).
	E:	The figures are bounded by straight lines.
	F:	The figures have a side of the same length.
	G:	The figures are bounded by lines (topological).
Vertex and Angle	D:	The figures have three vertices (triangle).
	C, I:	The figures have one right angle.
	D:	The sums of the angles of each of the figures is 180 degrees (triangle).
Others	H:	The plane figures are closed (topological).
	J:	The figures have the same area (equivalence).

Discussion of the responses

Because students are likely to characterize geometric figures according to relations between lengths of sides, to magnitudes of angles, and to shape (whether it is rectilinear or curved), we grouped the viewpoints accordingly in order to make it easy to reflect students' responses exactly. If all the viewpoints that are mentioned above appeared in the lesson, students would have acquired fundamental thinking patterns of geometry.

When this problem is used for evaluating learning, four aspects arise: the first is the number of positive responses (that is, the number of different figures proposed by students that have something in common with the given figure); the second is the mathematical quality of students' responses; the third is originality; and the fourth is the degree of elegance in students' expression of their thinking or ideas. Altogether, the evaluation sheds light on whether students can see daily phenomena in mathematical terms.

A Similar Problem: Making a Group of Related Figures

Substitute the rectangle below for the triangle in the previous problem.

ENLARGING FIGURES

KANJIRO KOBAYASHI

Elementary School at Chiba University

The Problem and Its Context

The problem

We want to enlarge this rectangle to double its dimensions. What drawing methods can you discover to do this? Draw your figure by as many different methods as you can. Explain your methods in words.

Pedagogical context

This problem is directly related to the topic of "scale drawing" that is prescribed for the sixth grade. There are several methods by which an enlarged figure can be drawn, such as drawing with a square lattice or basing the drawing on the length of a line and the angle measure. Even though several methods exist, we usually teach them one at a time.

The objective of the lesson is not just to help students understand the method of enlargement but to develop flexible and diverse ways of thinking about methods of drawing the required figure.

Looking at the problem from an open-ended perspective, we see that the process of drawing, not the result, is what is open in this problem, even though the required figure is uniquely determined. It is important, then, to encourage students to focus on the elements of the figures and the methods for drawing in order to discover a variety of methods.

The reason for selecting a rectangle for the original figure is that it has a suitable complexity of conditions for students at this level. A square seems too simple, since the enlargement is based only on the length of a side. A general triangle or quadrilateral is too complicated, since these figures have too many elements, such as several sides and angles, to consider.

Three class periods should be used for this lesson. The first period should focus on understanding the meaning and the basic property of "scale drawing." The second should focus on various methods of drawing. Discussion and a summary of the lesson should be the focus of the third period.

Expected Responses and Discussion of Them

Examples of expected responses

We grouped fourteen student responses into the following four categories. (Note. The author enumerated students' actual responses in the lesson, not necessarily the expected ones.)

- *Using sides and angles*

1. Extending and doubling the sides *BA*
   and *BC* to find the points *E* and *F* and
   drawing segments *EG* and *FG* perpendicular to *BE* and *BF*, respectively, to
   meet at the point *G*

$$2BA = BE, 2BC = BF$$
$$m \angle GEB = m \angle GFB = 90°$$

2. Extending and doubling the sides *BA*
   and *CD* to find *E* and *H*; extending and
   doubling the sides *BC* and *AD* to find *F*
   and *I*; then letting lines *EH* and *FI*
   intersect at the point *G*

$$2BC = BF = AI$$
$$2BA = BE = CH$$

3. Transferring the angle *a* formed by
   diagonals *AC* and *BD*, taking the
   point *O* as its vertex, and determining *OE*, *OF*, *OG*, and *OH* by
   the length of *AC*

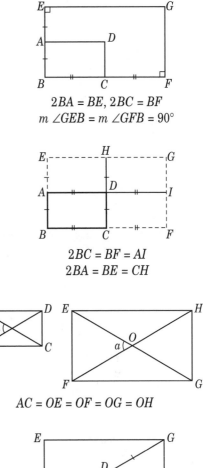

$$AC = OE = OF = OG = OH$$

4. Extending and doubling *BD* and *BC* to
   determine the points *G* and *F*, respectively, making a triangle *BGF*, and next
   a triangle *GBE* congruent to triangle
   BGF

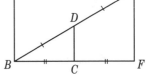

$$2BD = BG, 2BC = BF$$
$$\triangle BGF \cong \triangle GBE$$

5. Drawing lines *BP* and *RQ*, both through
   D, to make an angle equal to the angle
   made by diagonals *AC* and *BD*; determining the point *P* by extending and
   doubling *BD*, and the points *R* and *Q* by
   intersecting the extensions of *BA* and
   BC with the line *RQ*

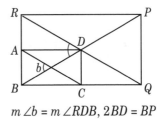

$$m \angle b = m \angle RDB, 2BD = BP$$

6. Finding the midpoints of segments AB, BC, CD, and DA and joining them as in the figure to find a point of intersection; determine the points I, J, K, and L as shown in the figure; drawing a quadrilateral by parallels through I, J, K, and L

- *Using square lattices*

7. Making square lattices of 1 cm in the given rectangle, and then doubling

- *Using the center of similarity*

8. Taking the point B as the center of similarity, and determining the points F, G, and E by $2\,BC = BF$, $2\,BD = BG$, and $2\,BA = BE$, respectively

9. Taking point 0, the intersection of the diagonals, as the center of similarity, and determining the points E, F, H, and G by $2\,OA = OE$, $2\,OB = OF$, $2\,OC = OH$, and $2\,OD = OG$

10. Taking one point on a side as the center of similarity

- *Other methods*

11. Arranging eight triangles that are congruent to triangle ABD

12. Arranging four rectangles that are congruent to $ABCD$

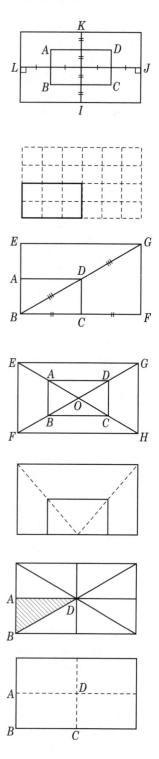

13. Drawing two circles with radii *CD* and *BD* and a common center *D*; determining the points *E*, *F*, and *H*, which are the intersections of the circles and extended lines *BA*, *BC*, and *CD*; extending *EH* to *G*; and joining points *E*, *G*, and *F* to produce a quadrilateral *EBFG*

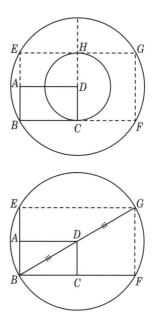

14. Drawing a circle with a radius *DB* and center *D*; joining the points of intersection of the circle and the extensions of *BA*, *BD*, and *BC* to produce a quadrilateral *EBFG*

Discussion of the responses

We grouped the methods into four sets; that is, drawing by measuring sides and angles, drawing by using square lattices, drawing by using the center of similarity, and other methods. The first group accounts for most of the students' responses.

Methods 1 and 2 above are based on doubling the lengths of sides. Method 1 is general; method 2 is devised by determining the point *G* as the intersection of two straight lines. Methods 3, 4, and 5 are based on diagonals and use the property that doubling the lengths of the diagonals doubles the lengths of the sides of the rectangle. Methods 3 and 5, which use the angle formed by two diagonals, seem to be difficult to draw. Though methods 4 and 8 look alike, students' underlying ideas are the differences between them, since the students' intention in method 4 is to produce a triangle whose sides are doubled in size. The idea in method 6 is unique. Method 7 is based on dividing the original rectangle into six equal parts.

Methods 8, 9, and 10 use a center of similarity and are relatively easy to discover. But only method 8 can be expected of students at the elementary school level. Methods 9 and 10, or a case taking a center outside the rectangle, seem too difficult for elementary school students to discover.

Methods 11 and 12, by arranging triangles or rectangles, use the idea of symmetry.

Methods 13 and 14 apply a property of a circle in the measuring or copying of sides or angles and are based on a higher-level idea. Method 13 is essentially the same as drawing by doubling the sides and diagonals. Method 14 is based on the property of an inscribed angle drawn to the endpoints of a diameter.

It is not always easy for the teacher to understand the ideas on which students base their drawings. This is mostly because students have difficulty expressing their ideas verbally. Thus, it is important that the teacher exercise patience with students. It is best that the teacher group students' similar responses when a variety of them are offered, but the teacher need not impose them on students.

PASCAL'S TRIANGLE

KOZO TSUBOTA

Fukazawa Elementary School, Setagaya Ward, Tokyo

The Problem and Its Context

The problem

See the table below and find as many rules or patterns as you can among the numbers.

```
1
1   1
1   2   1
1   3   3   1
1   4   6   4    1
1   5   10  10   5    1
1   6   15  20   15   6    1
1   7   21  35   35   21   7    1
1   8   28  56   70   56   28   8    1
1   9   36  84   126  126  84   36   9    1
1   10  45  120  210  252  210  120  45   10   1
-   -   -   -    -    -    -    -    -    -    -
```

Pedagogical context

Students have previously learned how to read graphs and tables as well as learned the properties of integers. As an application, students are asked in this problem to find as many rules as possible. The purpose is to encourage students to develop broader perspectives with respect to number tables.

The numbers are arranged in the shape of a right triangle. This helps the students to discover more rules, since it is easy for them visually to observe horizontally and vertically. The teacher must use care and encourage students to find rules by observing the table in several directions, not just horizontally and vertically. Students should be familiar with the phrase "rules or patterns among numbers." If not, the teacher should show one example to make the meaning clear.

The teacher first needs to help the students understand the problem, and then they should explore, trying to discover rules or patterns by trial and error. It is important for the teacher to allow sufficient time for this activity.

Expected Responses and Discussion of Them

Examples of expected responses

Several basic responses that elementary school students usually give are shown below. Later some more sophisticated ones are added. From the structure of the table, students can easily discover the following three rules:

1. All numbers in the left column are 1.
2. All numbers on the outside, right-down diagonal are 1.
3. Every number is the sum of (*a*) the number just above it and (*b*) the number to its left (upper left of the original number).

Other rules to be discovered are as follows:

4. The second column from the left is the sequence of natural numbers.
5. The third column from the left is the sequence of triangular numbers (see fig. 4.8).

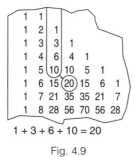

```
1
1  1
1  2 │ 1
1  3 │ 3 │ 1
1  4 │ 6 │ 4   1
1  5 │10 │10   5   1
1  6 │15 │20  15   6   1
1  7 │21 │35  35  21   7   1
1  8 │28 │56  70  56  28   8   1
```

```
 ·      ·      ·       ·
 1      3      6      10
```

Fig. 4.8 Triangular numbers

6. The sum of the first several consecutive numbers in any column is equal to the number located to the lower right of the last number (see figure 4.9).

```
1  1
1  2  1
1  3  3  1
1  4  6  4  1
1  5 (10) 10  5  1
1  6  15 (20) 15  6  1
1  7  21  35  35  21  7
1  8  28  56  70  56  28
```
1 + 3 + 6 + 10 = 20

Fig. 4.9

7. If we alternately first subtract and then add in a horizontal direction, the result is 0 (see fig. 4.10).

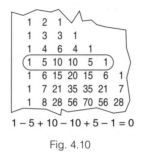

$$1 - 5 + 10 - 10 + 5 - 1 = 0$$

Fig. 4.10

8. The sum of all the numbers in a row is twice the sum of all the numbers in the previous row (see fig. 4.11).

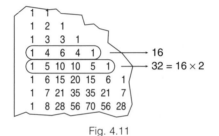

Fig. 4.11

9. The ratio of two consecutive numbers in a row is equal to the ratio of the number of numbers in the two parts of that row that are separated by an imaginary line or division between the two original consecutive numbers. (See fig. 4.12).

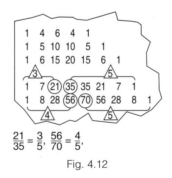

$$\frac{21}{35} = \frac{3}{5}, \frac{56}{70} = \frac{4}{5},$$

Fig. 4.12

10. The sums of numbers along diagonals drawn upward from the 1's in the left column form the "Fibonacci sequence" (1, 1, 2, 3, 5, 8, 13, ...), which is a sequence of numbers obtained by summing the preceding two terms, starting from 1 (see fig. 4.13).

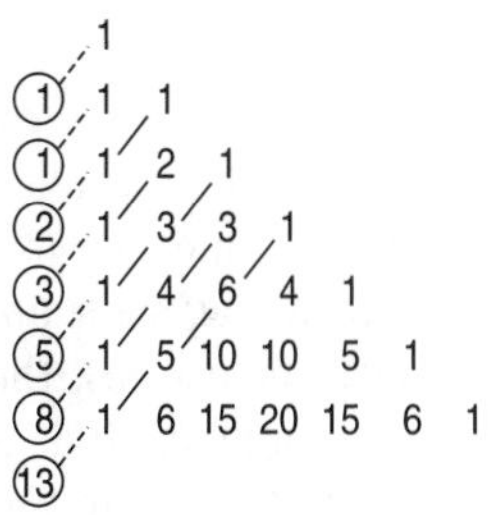

Fibonacci sequence: 1, 1, 2, 3, 5, 8, 13, ...

Fig. 4.13

The rules that the sixth-grade students actually found are shown in table 4.9.

Classification	Rule No.	Rule
Columns	1	The number of terms in a column decreases as we move to the right.
	2	All the terms in the first column are 1.
	3	In the second column from the left, we can see 1, 2, 3, ...
	4	Differences in the third column are 2, 3, 4, ...
	5	Differences in the fourth column are the numbers in the third column.
	6	The numbers in the nth column are the same as the numbers in the nth diagonal going down from the left.
	7	The sums of numbers mentioned in number 6 are equal to each other.
	8	Every number is larger than the number above it.
	9	(Expected response 6)
	10	Numbers in the seventh column are all multiples of 7, except for 1.
	11	The arrangement of column sums (except the right end) is symmetrical.
Rows	1	The arrangement of numbers in each row is symmetrical.
	2	Numbers at both ends are 1.
	3	(Expected response 8)
	4	(Expected response 3)
	5	The sum of the numbers in a row, except the first row, is always even.
	6	The number of numbers in rows increases by one.
	7	The numbers in the eighth row are multiples of 7, except 1 at both ends.
	8	The numbers in the middle of the rows are the largest.
Diagonals	1	When adding numbers diagonally downward from left to right, we get the number directly below the last one.
	2	The number of the numbers in successive lines parallel to the diagonal (formed by 1's) decreases by one.
	3	In the second line parallel to the diagonal, we can see 1, 2, 3, ...
	4	In the third line parallel to the diagonal, the differences are 2, 3, 4, ...
	5	In the seventh line parallel to the diagonal, the numbers are multiples of 7 (except at the beginning).

Table 4.9

Discussion of the responses

When elementary school students try to discover rules, and then express them verbally, their expressions cannot help but take a primitive form. In such instances, an effective strategy to help them group their ideas or rules; for example, grouping rules that concern columns, that concern rows, that concern diagonals, and so on.

When students are unable to discover some rules, the teacher should suggest some directions of thought, provided that doing so stimulates students' interest and ways of thinking.

Usually individual students present their rules to the whole class after discovering them. If there is insufficient time, the teacher should collect all notes from students at the end of the lesson in order to learn their individual responses, to check them, and then to share them later.

The rules in the list may be logically proved at the upper-secondary level as review. For example, expected response 7 can be proved as follows:

$$_nC_0 - {_nC_1} + {_nC_2} - \ldots + (-1)^n \, {_nC_n} = (1 + (-1))^n = 0$$

The Pascal's triangle problem, similar to the water-flask problem, can be used at the elementary, junior high, and senior high school levels. Further, students might be more interested in rules such as the Fibonacci sequence if they are shown photographs of flowers like daisies and sunflowers as concrete examples of the sequence in nature.

A Similar Problem

See the table on properties of multiplication in figure 6.1 in chapter 6.

INTERIM RESULTS OF BASEBALL GAMES

TETSUKO KUWABARA
Kamisugeta Elementary School, Yokohama

The Problem and Its Context

The problem

(At first, the teacher gives only table A to the students. Next, the teacher adjoins table B. After a while, the teacher adds table C.) Find as many rules or relations as you can from the chart in figure 4.14.

B Team	Games	Wins	Loses	Draws	Winning Ratio	Games Behind
X	50	34	15	1	0.694	
Y	46	25	17	4	0.595	5.5
Z	44	19	19	6	0.500	4.0
U	47	21	23	3	0.477	1.0
V	50	18	30	2	0.375	5.0
W	49	16	29	4	0.356	0.5

Fig. 4.14

Pedagogical context

The teacher poses the problem to students as follows:

- *Stage 1.* The teacher presents table A to the students and asks them to explore for relations or rules among the numbers. Students will be able to find some general rules or relations without referring to results of specific teams or wins, losses, and so on, since the names of the teams and the column heads are not yet given.

- *Stage 2.* the teacher gives the students table B. Students will now be able to deepen their understanding of how the outcomes of the baseball games are numerically represented in a table by relating what they have found to the number of games, wins, losses, and draws in this league. With added meaning, students will now be able to find more rules or relations than by having available the numerical table only.

- *Stage 3.* Finally, the teacher gives the students table C. Students will now be able to calculate the winning ratio and the number of games behind, if necessary by focusing their attention on the special case when the winning ratio is 0.500. They will see that it is convenient to use the winning ratio in comparison with the results of teams with different numbers of games and to use the number of games behind for expressing the relation between two teams.

Expected Responses and Discussion of Them

Expected responses

The list below shows the expected responses at each of the three stages of the lesson.

- Stage 1

    1. The students look across each row:

        $50 + 34 + 15 + 1 = 100, 50 \times 2 = 100, 34 + 15 + 1 = 50$
        $46 + 25 + 17 + 4 = 92, 46 \times 2 = 92, 25 + 17 + 4 = 46$

        Answers for the first and second calculations are always even, but those for the third calculation are odd or even. For example, $21 + 23 + 3 = 47$.

    2. The students look down each column:

        $34 + 25 + \cdots + 16 = 15 + 17 + \cdots + 29 = 133$

        (The sums of the left-most and right-most columns may also be considered, but they seem to be meaningless in finding a rule.)

    3. The students look along both rows and columns:

        $50 + 46 + \cdots + 49 = (34 + 25 + \cdots + 16) + (15 + 17 + \cdots + 29) + (1 + 4 + \cdots + 4)$

        This may give meaning to what seemed meaningless in (2) above.

- Stage 2

    1. (The number of games) = (the number of wins) + (the number of losses) + (the number of draws).

    2. Teams are arranged in order according to their position in the standings.

    3. (The number of games) + (the number of wins) + (the number of losses) + (the number of draws) is an even number.

    4. The sum of the number of games, the sum of the number of draws, and (the sum of the number of wins) + (the sum of the number of losses) are all even numbers.

    5. The sum of the number of wins equals the sum of the number of losses.

    6. (The sum of the number of games) = (the sum of the number of wins) + (the sum of the number of losses) + (the sum of the number of draws).

- Stage 3

    1. (The winning ratio) = (the number of wins) ÷ ((the number of games) – (the number of draws)) = (the number of wins) ÷ ((the number of wins) + (the number of losses)).

        (The winning ratio is independent of the number of draws.)

    2. (The number of games) = (the number of wins) ÷ (the winning ratio) + (the number of draws).

        (The number of games can be found by using the winning ratio.)

    3. (The winning ratio) = 1 – (the losing ratio).

(The winning ratio is always less than 1.)

4. Teams are listed in the table according to the order of their winning ratios.

5. The number of games behind has either 0 or 5 in its tenths place.

6. The number of games behind is a multiple of 0.5

7. (The number of games a team B is behind a team A)

= [(((the number of wins of A) − (the number of wins of B)) − ((the number of losses of A) − (the number of losses of B))] ÷ 2

= [(((the number of wins of A) − (the number of losses of A)) − ((the number of wins of B) − (the number of losses of B))] ÷ 2.

8. The number of games behind has an additive property; that is, the number of games a team C is behind team A is the sum of the number of games C is behind B and the number of games B is behind A (for example, the number of games Z is behind X is 5.5 + 4.0, that is, 9.5).

Discussion of the responses

1. *The relationship between responses in the first stage and responses in the second stage.*

By translating expression 1 in stage 2 into a formula using words, we may summarize in various ways:

- (The number of games) = (the number of wins) + (the number of losses) + (the number of draws).

- (The number of wins) = (the number of games) − (the number of losses) − (the number of draws).

- (The number of losses) = (the number of games) − (the number of wins) − (the number of draws).

- (The number of draws) = (the number of games) − (the number of wins) − (the number of losses).

Because most elementary school students do not realize that all the expressions above refer to the same idea, the teacher should help them to clearly understand it.

The rules along rows and columns (number 3 in stage 1) may be summarized as follows:

- (The sum of the number of games) = (the sum of the number of wins) + (the sum of the number of losses) + (the sum of the number of draws).

Furthermore, the rules along two columns (number 2 in stage 1), that is, 34 + 25 + ⋯ + 16 = 15 + 17 + ⋯ + 29, are examples of the rule in which the sum of the number of wins equals the sum of the number of losses. By careful consideration, students can understand the relations that they found when a concrete meaning is given. A discussion of parity may occur about the rule that the sum of the numbers of draws is even, and the teacher should draw students' attention to this.

In general, the teacher should help students to meaningfully interpret the rules they found in the first stage by connecting them to the expressions with words in the second stage in a concrete manner.

2. *Relationship among students' responses in the second stage. (Arrows mean "implies.")*

(The number of games) = (the number of wins) + (the number of losses) + (the number of draws).

(The number of games) + (the number of wins) + (the number of losses) + (the number of draws) is even because the first term is equal to the sum of the remaining three terms.

(The total number of games) = (the total number of wins) + (the total number of losses) + (the total number of draws).

The total number of games is even (because each game is counted twice).

(The total number of wins) = (the total number of losses). The total number of draws is even because each game is counted by both teams.

(The total number of wins) + (the total number of losses) is even.

(The total number of games) + (the total number of wins) + (the total number of losses) + (the total number of draws) is even.

3. *Relations mentioned in number 3.*

At first, the teacher should help students to understand that this "winning ratio" is different from an ordinary ratio to which they are accustomed. Students may jump to the conclusion that the winning ratio might be (the number of wins) ÷ (the number of games), as is customary. This conjecture is seen to be invalid by checking it by substituting actual values; it then becomes necessary to consider another approach. In such a case, observing the winning ratio to be 0.500 for a team would be helpful. Students will easily discover that the winning ratio has no connection with the number of draws.

Concerning the number of games behind, it will suffice to discuss the topic only briefly because its formal exploration requires some knowledge of positive and negative numbers. After identifying someone who succeeded in a correct formulation (or nearly so) through observation and individual work, the teacher may let students discuss their findings and then briefly summarize students' work.

A further development of this problem is possible; the following representation would be one possible topic. Assume that we are interested in comparing at some point during a season, the positions in the standings of each of six teams A, B, C, D, E, and F. By considering the winning ratio as a measure of relative strength, the situation may be represented by a directed graph (in graph theory); namely, we represent each team by a point in a plane and each game by segments joining pairs of teams (fig. 4.15). We attach arrows to segments in such a way that an arrow from A to B corresponds to A having more wins against B, and we attach to arrows the winning ratio of A. Thus when A has three wins and one loss against B, it may be represented by $A \xrightarrow{0.75} B$, in which 0.75 is the winning ratio of A against B. A and B having the same number of wins and losses may be represented by $A \xleftrightarrow{0.500} B$. This diagram would be helpful to globally understand the situation at that time.

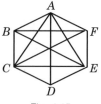

Fig. 4.15

Chapter 5

Examples of Teaching in the Lower Secondary Schools

I N THIS CHAPTER, we shall explain the main problems we tried in lower secondary schools in the format below. To save space, records of classroom teaching are included only for the typical examples. Where possible, similar problems are added for readers' information.

THE FORMAT OF EACH SECTION

The Problem and Its Context

1. *The problem:* The problem is stated in the way it was posed to the students.

2. *Pedagogical context:* The purpose of the problem is given, along with its relationship to the content in the textbook and to the mathematics program before and after the lesson.

Expected Responses and Discussion of Them

3. *Examples of expected responses:* Examples of expected responses to the problem are classified by viewpoints, including some high-quality responses that some students may make.

4. *Discussion of the responses:* Treatment for each type of response, the mathematical value of the responses, and how to evaluate the responses are detailed.

Record of the Classroom Teaching

5. *Teaching the lesson:* The process and steps used to teach the lesson, major questions, and related learning activities are discussed. (Minute details are omitted.)

6. *Remarks after the lesson:* Reflection on the teaching, the time needed, classroom discussion, collecting students' responses, and further development of the problem are presented.

REVIEW OF LINEAR FUNCTIONS

MASASHIRO HIROSHIMA

The Lower Secondary School at Fukui University

The Problem and Its Context

The problem

The graph and table in figure 5.1 show how values of two functions change. Figure 5.2 contains algebraic expressions of various functions.

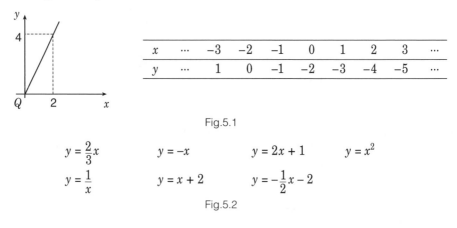

Fig.5.1

$$y = \frac{2}{3}x \qquad y = -x \qquad y = 2x + 1 \qquad y = x^2$$

$$y = \frac{1}{x} \qquad y = x + 2 \qquad y = -\frac{1}{2}x - 2$$

Fig.5.2

Choose functions from figure 5.2 that have a property in common with one or both of the functions illustrated in figure 5.1. Explain your point of view. Find as many viewpoints as you can.

Pedagogical context

This problem reviews the lesson on linear functions. Its purpose is to help students integrate what they have learned by making clear the relations among linear functions. We prepared this open-ended problem in a form in which teachers could easily present it and solicit students' responses.

In ordinary lessons on linear functions, the teaching style often involves asking the students to represent functions in the form of tables, graphs, or expressions. For example, "Represent the relation between x and y in a table, graph, Venn diagram, or expression" or "y is what function of x?" The subtopics are usually arranged in the following sequence: the meaning of linear functions, the range of variables, correspondence, the features of the graph, slope, intercept, and application. These topics are treated individually over a long span of time. Although such a step-by-step approach may be necessary at the stage of concept formation, a piecemeal understanding of subtopics does not necessarily guarantee an understanding of the whole. Students achieve such understanding only after they have a perspective from which to see the whole as a network of related components. When students engage in concrete activities for representing functions using tables, graphs, or expressions, they tend to direct too much attention to the technical aspects of the tables, graphs, or expressions. As a result, although it may be possible to develop basic knowledge or skills in isolated forms, the values of tables, graphs, or expressions will not be clearly appreciated, and we could not expect full and sure understanding of linear functions.

We prepared a problem in which students can compare the merits and demerits of each form of representation. After deciding which viewpoint to use, students can find some common feature among the different representations. They can then review what they have learned; appreciate the significance of each way of representation; understand the relationship between such concepts as set, correspondence, range, and order; and finally achieve a basic understanding of linear functions as a whole.

Expected Responses and Discussion of Them

Examples of expected responses

Viewpoints		Examples of Responses
Ratio of change	(1)	When x increases, y increases.
	(2)	The slope is the same.
	(3)	The rate of change is constant.
	(4)	The slope is positive.
	(5)	The graph is to the right and up.
	(6)	y is in proportion to x.
Expression	(7)	The function is represented in the form $y = ax$.
	(8)	y is a linear function of x.
Graph	(9)	The graph is a straight line.
	(10)	The graph goes through the origin.
Range	(11)	The range has no limitation.
Shape of graph	(12)	The graph is symmetric with respect to the origin.
Others	(13)	The graph goes through the first and third quadrants.
	(14)	The graph goes through the point (2, 4).

Table 5.1
Responses for the graph in figure 5.1

Viewpoints		Examples of Responses
Ratio of change	(1)	When x increases, y decreases.
	(2)	The slope is the same.
	(3)	The rate of change is constant.
	(4)	The slope is negative.
	(5)	The graph is to the right and down.
Expression	(6)	The function is represented in the form $y = ax + b.$ ($y = f(x)$)
	(7)	y is a sum of one being in proportion to x and the other being a constant number.
	(8)	y is a linear function of x.
Graph	(9)	The graph is a straight line.
	(10)	The y-intercept is the same.
	(11)	The y-intercept is negative.
	(12)	The graph goes through the point $(-2, 0)$.
Range	(13)	The range has no limitation.
Others	(14)	The graph goes through the point $(-1, -1)$.

Table 5.2
Responses for the table in figure 5.2

Discussion of the responses

In reviewing the lesson and what students have learned, along with their ways of thinking about the viewpoints listed above, the teacher should help students understand and appreciate the following points:

1. *Rate of change*

One point of view about the properties of functions is to observe "increase and decrease." Students must become aware that the rate of change in linear functions is always constant: when values of a function at x_1 and x_2 are expressed by $f(x_1)$ and $f(x_2)$, respectively, the average rate of change

$$\frac{f(x_2) - f(x_1)}{x_2 - x_1} = \frac{y_2 - y_1}{x_2 - x_1} = k$$

is constant in any interval. Therefore, it follows that for a linear function, a represents the slope of its graph and represents the change in y when x is increased by 1; when $a > 0$, the function is monotonically increasing; and when $a < 0$, it is monotonically decreasing. From these facts, students learn that the graph is a straight line.

If the teacher discusses only linear functions, students will not likely appreciate the fact that the average rate of change is constant. Therefore, the lesson should include other types of functions. Comparisons with other functions can highlight the features of linear functions.

2. *Algebraic expressions*

Representations of functional relations can be given by words, tables, graphs, and algebraic expressions. Almost all can be expressed in the form $y = f(x)$. Students have previously learned that linear functions can be expressed in the form $f(x) = ax + b$ or $y = ax + b$. Therefore, situations are needed in which the teacher can help students establish an abstract viewpoint by using algebraic expressions to compare functions expressed in different forms and then judge whether the functions are identical. At the same time, in developing students' mathematical thinking, it is important to integrate special cases into a general case. For example, the proportional relation $y = ax$ (which is learned in the seventh grade) is a special case of the linear function $y = ax + b$. Another example is to view $y - b$ as proportional to x in the function $y = ax + b$.

3. *Graphs*

In learning about functional relations by representing them as graphs, there is a tendency to focus solely on drawing graphs, writing expressions from graphs, or finding the value $f(a)$ when $x = a$. There is little opportunity to consider the merits of representing a function by a graph. So, the teacher should help students see that graphical representations of functions are important in visually understanding functional relations.

4. *Range*

In the eighth grade, the terms "domain" and "range" of a function are not used. It is important to firmly establish a way of thinking that focuses on domain and range. For example, what is the range of y if the domain of x is the set of integers, or the set of rational numbers? What if a linear function is considered? This method of thinking will enhance students' learning in the future when they begin to learn about inverse functions.

5. *Shapes of graphs*

Viewing the graph of a function as a geometric figure helps students understand the features of the function. Observing the graph's symmetry or position in relation to the graph of another function can illuminate the differences between the functions. This is especially true when students learn about the functions $y = ax^2$ and $y = ax^3$. Focusing on symmetry will enable students to understand the main features of each function.

After summarizing what has been learned about these functions, including what appeared to be ambiguous or "messy" at first, it is useful to consider the relation between functions and equations. Students usually regard x and y in a linear equation $ax + by = c$ as unknown and unchanging numbers. However, if x and y are considered as variables, then $ax + by = c$ can be seen as a function with values that are true or false, and we have a set of solutions as a graph. Regarding an equation as a function or a function as an equation supplements earlier learning. For example, lessons on simultaneous linear equations in two variables are often confined to the case in which one, and only one, solution exists. It would be effective to go further and consider the cases in which a solution does not exist or when the solution is not unique and has infinitely many solutions. Students can be visually helped to understand these cases as a background for indeterminate or inconsistent solutions.

Record of the Classroom Teaching

Teaching the lesson

The purpose of this lesson was to summarize students' learning of linear functions.

Teacher's Instructions	Learning Activities	Minutes Allocated
1. Pose the problem and explain it with work sheets.		5
2. Let's select functions from figure 5.2 that have properties in common with the graph in figure 5.1 from many points of view.	Write responses on the worksheet. Example 1. When x increases, y also increases. Example 2. and so on	10
3. For the table in figure 5.1, let's select functions that have the same properties.	Write responses on the worksheet.	10
4. Discuss with others in your group what you have observed. Write down in red what you newly discover in the discussion.	Discussion in groups of four students.	20
5. Summarize what you have observed in your group.	Leaders of each group write their summary on a transparency. (Since the time was short, the work was left to each group for the next lesson.)	5
6. Let's present what each group has observed.	Leaders present their groups' ideas using the overhead projector.	20
7. Let's confirm what was presented. Let's put them in the order of viewpoints.	The teacher summarizes students' presentations and writes them on the chalkboard.	30

Table 5.3
Lesson plan

Viewpoints		Responses	Number of Positive Responses	
			Group	Individual
Rate of change	(1)	When x increases, y also increases.	1	1
	(2)	The slope is the same.	11	38
	(3)	The rate of change is constant.	3	5
	(4)	The slope is positive.	4	6
	(5)	The graph is right and up.	11	20
	(6)	y is in proportion to x.	5*	8
Representation by expression	(7)	The expression is represented by $y = ax$.	1	2
	(8)	y is a linear function of x.	7	13
Representation by graph	(9)	The graph is a straight line.	8	10
	(10)	The graph goes through the origin.	11	41
Range	(11)	The range has no limitation.	2*	2
Shape of graph	(12)	The shape is symmetric with respect to the origin.	1	2
Others	(13)	The graph goes through the first and third quadrants.	1	3
	(14)	The graph goes through the point (2, 4).	8	17
	(15)	It is continuous.	0	5

Note: The total number of students was 44, and the number of groups was 11.
Individual Response means facts written on the worksheet at stage 3 of the lesson plan, and Group
Response those at stage 4, among which * means the number of facts discovered only through group work.

Table 5.4
Summary of students' responses for the graph in figure 5.1

Viewpoints		Responses	Number of Positive Responses	
			Group	Individual
Rate of change	(1)	When x increases, y decreases.	3	3
	(2)	The slope is the same.	11	34
	(3)	The rate of change is constant.	3	8
	(4)	The slope is negative.	2	3
	(5)	The graph is to the right and down.	10	21
Representation by expression	(6)	The expression is represented in the form $y = ax + b$.	1	1
	(7)	y is in proportion to x and a constant number.	0	5
	(8)	y is a linear function of x.	6*	9
Representation by graph	(9)	The graph is a straight line.	5	8
	(10)	The y-intercept is the same.	11	40
	(11)	The y-intercept is negative.	2	3
	(12)	The graph goes through the point $(-2, 0)$.	8*	10
Range	(13)	The range has no limitation.	2*	2
Other	(14)	The graph goes through the point $(-1, -1)$.	3	3

* This means the number of facts discovered only in group work.

Table 5.5
Summary of students' responses for the table in figure 5.1

Remarks after the lesson

1. Students expressed a variety of points of view, but most of their expressions were not of high quality. Students lacked an ability to express their ideas clearly.
2. Few pupils focused on the range or the shape of the graph.
3. Students focused on intuitive features such as the graph having the same slope, going right-up or right-down, passing through the origin, having the same y-intercept, or being a straight line.

These observations reveal two approaches that students took to the problem. In the first approach, students represent each function by a graph, thereby grasping the larger picture, and then consider differences between functions locally. In the second approach, students represent each function by an algebraic expression and then consider differences between them locally. In neither case are students aware of the significance of representations by a graph or an algebraic expression.

This lesson helped students integrate what had previously been isolated or ambiguous to them into a more coherent whole.

Similar Problems

Properties of Functions (1)

The original problem can be modified into the following problems.
Examine the representations of functions 1–8 and answer the questions that follow.

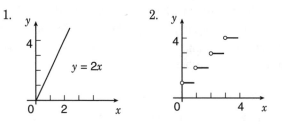

1. $y = 2x$

2. (graph)

3.

x	−4	−3	−2	−1	0	1	2
y	2	1.5	1	0.5	0	−0.5	−1

4.

x	−2	−1	0	1	2	3	4
y	8	7	6	5	4	3	2

5.

x	−3	−2	−1	0	1	2	3
y	3	2	1	0	1	2	3

6. Let y be the residue of x divided by 3, and
$x \in X, X = \{0, 1, 2, 3, 4, 5\}$
$y \in Y, Y = \{0, 1, 2\}$

7. $y = \dfrac{6}{x}$

8. $y = x^2$

Question 1. Find the properties of function 1 that are common to at least two of the functions in 2–8.

Question 2. Find the properties that are common to at least two of the functions 2–8.

Question 3. What have you learned in today's lesson?

Properties of Functions (2)

Question 1. (A) Draw the graphs of the following functions. (The domain is the set of real numbers.)

1. $y = \dfrac{2}{5}x$ 2. $y = \dfrac{2}{5}x^2$ 3. $y = \dfrac{2}{5}x^3$

4. $y = -\dfrac{2}{5}x$ 5. $y = -\dfrac{2}{5}x^2$ 6. $y = -\dfrac{2}{5}x^3$

(B) Find as many properties as you can that are common to at least two of the functions in 1–6.

Question 2. Answer the following questions about the graphs of the functions 1–5.

1. $y = x$ 2. $y = x^2$ 3. $y = x^3$ 4. $y = x^4$ 5. $y = x^5$

a. Compare the values of y when $x = 2$ in the functions 1–5.

b. Study how the values of y change when the values of x change from 2.

c. For functions 1–5, find as many common aspects as possible. Also find as many aspects as possible in which they differ.

Points of Intersection of Two Intersecting Circles

Figure 5.3 shows six sets of intersecting circles. In each set, the circles O and O' intersect at the points A and B, and a straight line passing through A intersects the circles at the points P and Q, respectively. Similarly, a straight line passing through B intersects circles O and O' at the points R and S, respectively.

Among the relations, properties, or procedures of proof that hold in figure 5.3a, select those that hold in some of the figures 5.3b–5.3f. State the bases for your selections.

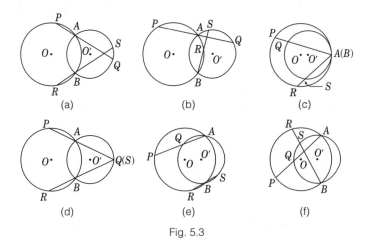

Fig. 5.3

RIGHT TRIANGLE AND
THE "CONNECTING MIDDLE POINTS" THEOREM

HIROSHI ISHIYAMA
The Lower Secondary School at Yamagata University

The theorem of connecting middle points states that *the segment joining the midpoints of two sides of a triangle is parallel to the third side and equal to one-half its length.*

The Problem and Its Context

The problem

In a triangle ABC, angle B is a right angle, and point D is the midpoint of the opposite side AC.

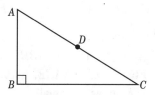

Question 1. Read the following sentences and draw the figure.

Let E, F, G be the midpoints of line segments AD, DC, and BC, respectively.
Let H be the point on AB such that $DB \parallel EH$. Join B and D, E and H, F and G, H and G, respectively, by lines.

Question 2. In the figure you just drew, find as many relations as you can and write them in the form of if-then statements. The following categories may help you find relations:

(1) side (2) shape (3) angle (4) others

Question 3. Do the relations you found in question 2 also hold true in the case where angle B is not a right angle?

Pedagogical context

In teaching geometry, it is important not only to help students understand the basic properties of geometrical figures but also to help them consider geometrical figures in a developmental and unified way. Students also need to develop the ability to think deductively as well as intuitively.

Some important aims of teaching the properties of similar figures in the eighth grade are to help students understand the meaning of geometrical similarity, the relation between similar triangles and measurement, and the practical applications of similar figures.

The specific purposes of this problem are to help students understand the importance of the "connecting middle points" theorem and its converse, develop their ability to devise proofs using if-then statements, and understand the logical relations among properties of similar triangles.

This lesson was positioned in the eighth-grade unit on similar figures between the topics "triangles and proportion" and "the center of gravity of triangles." It builds on the intuitive treatment of similarity learned in elementary school and geometrical congruence learned in the seventh grade.

The eighth-grade topic "triangles and proportion" includes the theorem of connecting middle points. This lesson explores the converse of the theorem.

Expected Responses and Discussion of Them

Examples of expected responses

Relations between sides: (The reasons are omitted here.)

$$FG \mathbin{\#} \frac{1}{2} BD \qquad AH = BH \qquad EH \mathbin{\#} \frac{1}{2} BD \qquad GH \mathbin{\#} \frac{1}{2} AC$$

(The symbol $\#$ means "is equal and parallel to.")

    *$AD = CD = BD \ (= HG) \quad$ *$HE = AE = ED = DF = FG = CF$

Relations between sizes and shapes:

    $\triangle CGF \sim \triangle CBD \qquad \triangle AHE \sim \triangle ABD \qquad \triangle BHG \sim \triangle BAC$

    Isosceles triangles * ($\triangle FGC, \triangle DBC, \triangle AHE, \triangle ABD$)

    $\triangle DBC = \triangle ABD = \square HGFE$ (areas are equal)

    $\triangle AHE = \triangle FGC = \frac{1}{4} \triangle ABD = \frac{1}{4} \triangle BDC$

    $\triangle BHG = \frac{1}{4} \triangle ABC$

Others:

    HGFE is a parallelogram.

    Point D is the circumcenter of triangle ABC.*

    Triangle HBG is a right triangle.*

(*Note:* Relations of angles are omitted because they were treated in similarity;
       *denotes those that do not hold when the angle B is not a right angle.)

Discussion of the responses

The expected responses to this problem will appear only after students draw the figures correctly and understand the underlying assumptions. In the lesson, the teacher should have students draw figures correctly by using a right angle square and a compass. After checking their work themselves, they should begin to look for valid relations.

The teacher should consider the following three points:

1. *The starting point in studying the relations*

    The students have previously learned the "connecting middle points" theorem. Its proof was treated when students studied similar triangles. However, if the students can find and then explain logically the converse of the theorem (that is, if $AE = DE$

and $\overline{EH} \parallel \overline{BD}$, then $AH = BH$), many other relations follow. Understanding the converse will clarify the logical relations among the many properties.

The sequence of the responses is shown in the flowchart below.

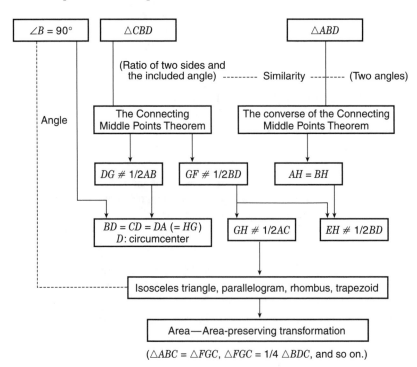

($\triangle ABC = \triangle FGC$, $\triangle FGC = 1/4 \triangle BDC$, and so on.)

2. Propositions when the triangle is not a right triangle

Conventional teaching generally leads students "from general to specific." However, this lesson can teach generalization by considering the relations that hold even when angle B is not a right angle. Even when there seems to be no relation, it is possible to find many common properties. It is useful to have the students conjecture the result before drawing figures concretely, and then check their conjectures from the figures they drew. The relations shown in the diagram above will then become clear, and students will more deeply understand the importance of the theorem.

3. Important points of the theorem

Ordinarily, theorems are used as tools in developing proofs; however, this tends to be too formal for the students. To acquire the ability to explore actively, they must appreciate both the meaning and the significance of the theorem. The "connecting middle points" theorem has been one of the most important theorems taught in junior high schools. This lesson helps to understand why this is so.

Record of the Classroom Teaching

Teaching the lesson

Teacher	Student	Minutes
1. Pose the problem (on worksheet); clarify purpose of lesson.	1. Draw the figures correctly. 2. Find as many properties as you can and write them down.	5

2. Have students—
  a. draw the figures with right-angle square and compass;
  b. make their assumptions clear;
  c. check their work.
3. Give guidance (to "slow" students).
	3. Read the problem.
	4. Draw figures correctly and quickly.
	5. Mark those parts that are assumed to be equal in the given problem.
	6. Criticize each other's drawings and confirm whether they are correct.
		5

4. Have students—
  a. examine the figures;
  b. find the valid relations;
  c. express them symbolically.
	7. Look at the figures and find the relations among sides, and write them down.
	8. Find the relations among shapes and sizes, and write them down.
		15

5. Have students present the relations they have found; write down the number of responses.	9. Raise your hands if you found the same property; indicate what you felt was difficult.	5
6. Discuss common properties with students.... Make the reasons clear and classify the properties according to the viewpoints used to find the relations.	10. Present your findings while briefly giving the reasons.	15
7. Summarize the learning making use of students' starting points. 8. Reconfirm the importance of the theorem (singular). 9. Help students understand logical relations between the theorems.	11. Summarize today's learning... by reflecting how you proceeded to the solution from attempts that failed at the first stage.	3
10. Have students conjecture the relations in the case where angle B is not a right angle. 11. Check the responses. 12. Assign part of the lesson as homework.	12. Let's check the case where angle B is not a right angle from the summary.	2

The lesson was tried according to the following plan:

Step	Mins.
1. Distribute a worksheet that includes the original (given) drawing and the instructions for adding points and lines. Explain the purpose of the lesson. Clarify the written instructions as necessary.	5
2. Suggest that the students use a right-angle square and compass to draw the points and lines precisely, and mark the parts that are "equal by assumption." Have them check one another's work to make sure that the drawings are correct.	5
3. Instruct the students to find as many relations as they can in the figure and to write them down in symbolic form. Tell them to look first for relations among sides and then for relations among shapes and sizes.	15
4. Have the students present the relations they found to the class and to briefly explain their reasoning. Tell those who are listening to raise their hands when they hear another student name a relation that they found. Record the number of relations that each student found. Ask them to express what was difficult for them.	5
5. Discuss the relations with the students, classifying the relations according to the viewpoints described in the section "Examples of Expected Responses."	15
6. Ask students to reflect and express how they proceeded from "difficulty" to understanding in the exercise.	?
7. Reiterate the importance of the "connecting middle points" theorem and summarize the logical relations between the theorem and its converse.	?
8. The case where angle B is not a right angle can be introduced and assigned as homework.	

Summary of students' responses

The number of responses made by individual students ranged from two to more than thirteen. The table below shows the distribution of the number of responses among forty-one students.

Number of responses	0,1	2,3	4,5	6,7	8,9	10,11	12,13	over 13
Number of students	0	4	8	10	8	6	4	1

The table shows that many students did not have as many responses as expected, perhaps because it took a long time for them to reason why $AH = BH$. Students who found that $AH = BH$ and $FG \# 1/2DB$ in the early stage were among the group of middle achievers, while those in the high group had a relatively small number of responses. Perhaps this is due to their tendency to pay more attention to the accuracy of reasoning or to the equivalent relations among their findings.

- Most students who had two to five responses stuck too much to the starting point $AH = BH$. However, during the discussion of their presentations, they developed others and also appreciated the importance of the converse of the "connecting middle points" theorem.
- Many students in the high-achieving group had a smaller number of responses than expected, perhaps because they summarized equivalent relations into one type. They were more successful in logically relating properties of geometrical figures than expected.
- In the homework for the case where angle B is not a right angle, most responses were related to the responses for the case where angle B is a right angle. The reason stated was that a difference was only a partial one. Students commented that a solution was possible by comparing accurate drawings.
- Two students found that point D is the circumcenter of triangle ABC when angle B is a right angle. They checked it by drawing figures using a compass.
- Several students modified the given situation. They drew auxiliary lines or labeled the point of intersection of HG and BD, and then tried to find some properties. However, after discussion, we concluded that making conclusions from such modifications was not appropriate for this lesson.
- Some students made flow charts as their summary.
- In the homework for the case where angle B is not a right angle, many students distinguished between assumptions and conclusions by drawing figures using different colors, and then explained their results using the figures. This work was clear evidence that the viewpoints they used in thinking about the problem were improved. (See figures below.)

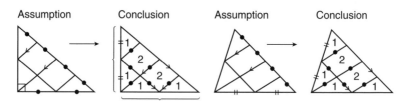

Assumption   Conclusion   Assumption   Conclusion

Remarks after the lesson

In teaching about geometrical figures, there is a strong tendency to ask students just to state theorems, supplemented by exercises in proof after analyzing what is given in the assumptions and the conclusion. However, to meet our objectives, students need to read the problem carefully, draw the figures themselves, start by conjecturing, use reasoning, and think flexibly. This approach makes it possible for students to experience broad and deep learning, by paying attention to both structural and personal ways of acquiring knowledge and understanding.

The implications for future teaching are given below:

1. It would be more desirable to use this problem after the topic of "center of gravity of a triangle," as a summary stage in the unit of "similar figures." Then students could

understand the logical relations among various properties by applying the properties of geometrical figures or the theorems in a variety of cases.

2. The case where angle B is not a right angle should be taught in another lesson. Expecting the students to see the properties of geometrical figures *according to the kinds of triangles* was too much to reasonably expect in a one-hour lesson.

3. It would have been better to introduce group discussion of students' viewpoints at the stage of drawing conclusions after students individually drew figures. This might have evoked higher-quality responses, although it would have taken more time than was scheduled.

4. A flowchart of the logical relations among the properties helps to summarize students' learning. Related problems could help broaden and deepen students' thinking about geometrical figures.

5. One student said, "At first I couldn't find what was required of me, but I found many properties easily after I found the starting point $(AH = BH)$. It was fun." Another said, "I was impressed by the good points of applying the theorem once again." These comments show that this problem has merit in stimulating students to think more deeply.

6. In the interview after the lesson, students commented that—

   • Drawing figures correctly and making assumptions clear are useful for later learning.

   • The theorems not only should be memorized by using figures, algebraic expressions, or verbal expressions, but can be appreciated by applying them.

   • Although a larger number of responses from students is good, it is important to summarize and categorize equivalent responses.

   • From a mathematical point of view, it is interesting to find that a variety of approaches is possible in attacking one problem.

Similar Problems

Parallelogram and Angle Bisector

   ABCD is a parallelogram, and *BE* is the bisector of angle *B*. *BE* intersects line *CD* at point *F*.

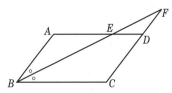

   Question 1. Draw the bisector of the exterior angle of *ABC*. This bisector intersects line *DA* at point *G*, and line *DC* at point *H*. Draw this figure.

   Question 2. What can you discover about the following from these two figures?
   (1) relations   (2) shape   (3) others

   Question 3. Transform *ABCD* to a rectangle. Do the results you found in Question 2 change?

A Quadrangle Inscribed in a Circle

   Examine figures 1 and 2 below. Find as many relations as you can in the figures. Try to prove that the relations you found are true.

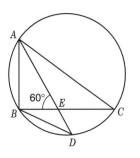

Fig. 1

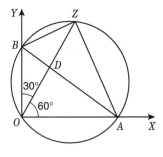

Fig. 2

NEW OPERATIONS

MASAKAZU AOYAGI

Faculty of Education, Chiba University

The Problem and Its Context

The problem*

An *operation* produces a new number from two given numbers. Addition, subtraction, multiplication, and division are the four basic operations. Find as many new operations as you can.

Pedagogical context

Topics such as this are taught as part of the unit on number and algebraic expressions. By focusing on the structure of a set of numbers,* they give meaning to the concepts of (a) a set being "closed" with respect to some operation; (b) the commutative, associative, and distributive laws holding with respect to an operation; and (c) the identity and inverse elements. The objectives in teaching such topics are to help students—

- reconsider the basic operations of addition, subtraction, multiplication, and division in the sets of natural numbers, integers, and rational numbers;
- integrate their knowledge about operations and sets; and
- become acquainted with algebraic structures.

In their lessons, teachers sometimes use examples in which some of the three laws do not hold, or some operations are not closed, in order to help students understand the validity of the three laws or the meaning of a set being closed. It is natural for students to cite examples of operations they have learned in the set of natural numbers that are not commutative (i.e., subtraction or division) or that are not closed. Moreover, to teach the meaning of a binary operation, it is helpful to give examples of operations different from the four operations and to have students think about the structures of the number sets on which these new operations are defined.

After the structures of the sets of natural numbers, integers, and rational numbers are taught, this problem can be used as an introduction to the unit on new operations.

Expected Responses and Discussion of Them

Examples of expected responses

We expect that many responses will be of the form $a \bigcirc b = la + mb + n$, in which $\bigcirc$ represents an operation and l, m, and n are constant numbers.

Others will be in the form of $a \bigcirc b = a^2 + b^2$, or $a \bigcirc b = |a \pm b|$. Also, some students

---

* Translator's note. This problem was used in the 1970's. From the 1980's onward, most of the topics mentioned here are no longer in the textbooks.

may invent an operation that produces the greatest common divisor or the least common multiple of two numbers.

Discussion of the responses

The following examples show how students' responses can be used to teach structures with respect to the operations in number sets. These examples are of the form $la + mb + n$, where l, m, and n are special constant numbers.

1. *Closure*

   Have students consider whether the set of integers is closed with respect to the operation $a \bigcirc b = (a + b) \div 2$.

2. *Commutative law, associative law*

   Have students consider whether the commutative and associative laws hold in the set of integers with respect to the following operations:

   A. $a \bigcirc b = a + b + 1$
   B. $a \bigcirc b = 2(a + b)$
   C. $a \bigcirc b = a$
   D. $a \bigcirc b = 2a + b$

The table below indicates whether the commutative and associative laws hold for each operation in this example. "T" (true) indicates that the law holds; "F" (false) indicates that it does not.

	A	B	C	D
Commutative law	T	T	F	F
Associative law	T	F	T	F

Students can use this table to classify their responses into several types. To extend the problem, you can ask them to invent some new operations that belong to each type.

Record of the Classroom Teaching

Teaching the lesson

This lesson was taught in the first lesson in the unit "New Operations." Students had finished learning the structure of the sets of natural numbers, integers, and rational numbers before this teaching took place. They understood the terms *closure, commutative law, associative law, distributive law, identity element,* and *inverse element.*

The teacher distributed the handout shown in figure 5.4 to the students.

The Eighth Grade Class ____ Number ____ Name _____	
Let's determine many other operations by referring to the examples below.	
Expression	Rules (in words)
Example 1 1. $1 \oplus 2 = 3$ 2. $2 \oplus 3 = 0$ 3. $4 \oplus 3 = 2$	This is the same as the usual addition when the sum is 4 or less. The answer is the sum minus 5 when the sum is more than or equal to 5.
Example 2 1. $3 * 4 = 10$ 2. $5 * 1 = 11$ 3. $7 * 3 = \square$	

① 1.
    2.
    3.	

② 1.
    2.
    ⋮	

Fig. 5.4

Table 5.5 shows the progress of the teaching.

Students' Activities and Related Mathematical Content
1. The students discuss Example 1 on the worksheet. They think about computation using clock arithmetic (i.e., the hand makes a complete revolution every five hours). They define a new operation $\oplus$ such that $2 \oplus 1$ means one hour after 2 o'clock (i.e., 3 o'clock).   They compute the following with respect to the new operation $\oplus$. $\qquad 2 \oplus 1 \quad 2 \oplus 2 \quad 2 \oplus 3 \quad 2 \oplus 4$ (10 minutes)
2. Looking at Example 2 on the worksheet, the students determine the rule for the operation * in the following: $\qquad 3 * 4 = 10 \qquad 5 * 1 = 11$ Using this rule, students find the answer to $7 * 3 = \square$. They then consider how * is similar to or different from the four operations.   (20 minutes)

3.	The students are asked to invent other new operations and write them down on their worksheets using the symbol ○ to denote each new operation.
	(30 minutes)
4.	The students make an operation table for the operation ⊕ in Example 1 and use the table to examine the operation's algebraic structure. Is the set {0, 1, 2, 3, 4} whose elements are the numbers on the clock closed with respect to the operation ⊕? What is the identity element? What are the inverse elements? Do the various laws hold in this set?
	(40 minutes)
5.	Summary Many operations exist other than the four operations. There are operations that have the same structures as the four operations, though they look different.
	(45 minutes)

Table 5.5

In the following, "T" means the teacher, and "S" means the students.

T: Let's learn some new operations today.
   Look at 2 + 3. What is this?

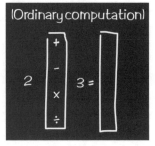

S: (astonished) 5.
T: Yes. Subtract!
S: (loudly) –1.
T: Yes. Multiply!
S: (all students at once) 6.
T: Yes. Divide!
S: (slowly) 2/3.
T: There are two numbers, 2 and 3. An operation means that from the two given numbers, we get one number according to a certain rule. Right? So,

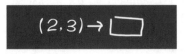

Today, we will think about an operation that may appear a bit strange. (The teacher draws a picture of a clock on the chalkboard.)
On this clock, the hand makes a complete revolution every five hours. If the hand is on zero, it will be back on zero after five hours. Imagine that you go to some country where this clock is used.
Suppose it is two o'clock. After one hour?

S: Three o'clock.
T: It is two o'clock now. After two hours?
S: (all in unison) Four o'clock.
T: It is two o'clock. After three hours?

S: Twelve o'clock. (a stir and laughter)

T: There is no twelve on this clock. Use only the numbers on the clock in this country.

S: Zero o'clock. (A chorus of "OK!")

T: On this strange clock, the operation looks like addition, but it is different. What is the difference?

S: It is the same up to four. If it is over four, we subtract five, and we take the rest.

T: Yes, that is right. In this computation, we use the symbol ⊕ to distinguish it from addition.

How about the following computation?

$$2 \oplus 1 = \quad 2 \oplus 2 = \quad 2 \oplus 3 = \quad 2 \oplus 4 =$$

The students write their answers in their notebooks. The teacher goes around the classroom, observing and giving guidance individually. Errors are found for the last two operations.

T: Now look at the Example 2 on the handout sheet. Do you know what the symbol * means? Discuss it in your groups.

Looking at the worksheet, the students discuss the meaning of * in each group.

T: Well, do you have any guess? (Almost all the students raised their hands. The teacher called on one student.)

S: Double the first number, and add the second number.

T: Yes. What is the number in □?

S: 17.

T: Suppose we represent two numbers by a and b, then what is the answer?

S: (Almost all of the students raised their hands.) $2a + b$. ("OK!")

T: That's right.

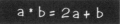

$$a * b = 2a + b$$

The operations ⊕ and * are a bit different from the four operations. In what ways are they the same? Think about it.

S: Using two given numbers, we make one number according to some fixed rule. In this way they are the same, I think.

T: That's right. Speaking exactly,

An operation makes a new number by joining two numbers a and b in a set, according to some fixed rule.

The word "operation" can be used not only for the four operations, but also for new and different computations.

Let's proceed. Look at the rows ①–⑤ on the worksheet. Write down as many new operations as you can. Use the symbol ∘ to represent the way to join two numbers. If this seems awkward, an "arrow" symbol may also be used.

$$2 \circ 3 = 10, \ (2, 3) \rightarrow 10$$

The students write for a short time.

T: H, explain your ideas.

S: Multiply by 2 and subtract. Multiply by 3 and add.

T: Explain more concretely, for example, by using algebraic expressions.

S: $1 \circ 2 = 0$ and $1 \circ 1 = 4$.

T: Using the two numbers given, you get a new number.
Are there any other examples?

S: Subtract b from a, then subtract it from 5.

T: OK. It's good. It seems that there are other examples, but I will see them when I look at your handouts. So, let's proceed.

Then the teacher progressed to item 4 in table 5.5.

- The class completed the operation table for the computation on the clock. (A blank table was printed on the back of the handout.) Students examined the operation table, looking for the rules. They examined "closure," "identity element," "inverse element," and "commutative law" from the symmetry with respect to the diagonal line of the table.

- After comparing the new operation to the four basic operations, they summarized the lesson.

Examples of students' responses

Table 5.6 lists the students' responses and the frequency of each response. The number of students is 42.

Response	Frequency
(i) $a \circ b = la \pm mb \pm n$	89
$la \pm mb$	54
$a/l \pm b/m \;\; (l \neq 0, m \neq 0)$	20
$la \pm mb \pm n \; (n \neq 0)$	15
(ii) $a \circ b = a\,(la \pm mb) \pm n$	15
(iii) $a \circ b = ab - b$	5
(iv) Square one number or both numbers, and then add or subtract them.	8
(v) Arithmetic on the clock.	10
(vi) Others	9
Represent the answer in the binary system.	3
Greatest common divisor or least common multiple of a and b.	2
Absolute value as in $|a| + |b|$, $|a + b|$, etc.	2
Find the smallest divisor other than 1 of each number, and then add them.	1
If $a - b \geq 1$, then $a - b$; and, if $a - b < 1$, then $a - b + 5$.	1

Table 5.6

Remarks after the lesson

1. In this class, the group discussions were very active, and students presented more responses than expected.

2. After arranging the responses, the teacher can present *them* to the class at the beginning of the next lesson, and proceed using those operations. This practice will likely generate more student interest in learning new operations.

3. In this lesson, no more than ten of forty-five minutes were used to pose the open-ended problem, and students then thought about the problem in a free and open manner. The teacher treated the operation on the clock as an introduction in order to help students understand the meaning of the open-ended problem. If students are prepared in this way, they can give many responses in a short time.

SOME PROPERTIES OF A NEW OPERATION

WATARU MIYASATO
Hamawaki Lower Secondary School, Beppu City

The Problem and Its Context

The problem

1. Prepare two disks A and B of different sizes. Write 0, 1, 2, 3, and 4 on their perimeters in equal distances, respectively, as in the picture on the left below. Join them at their center by an eyelet so that each can be rotated freely around the center.

2. A new addition operation $3 \oplus 4$ is defined using the disks; that is, the addition denoted by $3 \oplus 4$ can be found as follows: when the 0 on disk B is aligned with the 3 on disk A, the sum is the reading on A that aligns with the 4 on B (see picture on the right below).
   Find as many rules for this operation $\oplus$ as you can with respect to the set
   S = { 0, 1, 2, 3, 4 }.

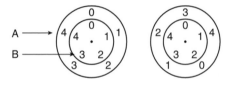

Pedagogical context

Students are expected to find properties of the new operation $\oplus$, which is defined on the set S of numbers written on the disks. The purpose is to have students understand that the set S, which has only five elements, has the same algebraic structure with respect to $\oplus$ as the set of rational numbers has with respect to ordinary addition.

Expected Responses and Discussion of Them

Examples of expected responses

1. *About the operation*

   a. The operation can be summarized as in the table below.

$\oplus$	0	1	2	3	4
0	0	1	2	3	4
1	1	2	3	4	0
2	2	3	4	0	1
3	3	4	0	1	2
4	4	0	1	2	3

   b. When computing with $\oplus$, if the sum of two numbers is less than 5, the result is the same as for ordinary addition; if the sum of two numbers is not less than 5, we subtract 5 from the sum.

c. In the table the same numbers lie along each right-up diagonal.

d. The numbers are symmetrical with respect to the main right-down diagonal (i.e., the commutative law).

$\oplus$	0	1	2	3	4
0	0	1	2	3	4
1	1	2	3	4	0
2	2	3	4	0	1
3	3	4	0	1	2
4	4	0	1	2	3

e. In each row and in each column, each number 0–4 appears exactly once.

2. *About the structure of the set of numbers on the disks*

With respect to the operation $\oplus$ on the set S = {0, 1, 2, 3, 4} (the numbers on the disk), the following properties can be found:

a. Since the results of the computation are also elements of the set S, the set S is closed with respect to the operation $\oplus$.

b. Since $3 \oplus 4 = 2$, and $4 \oplus 3 = 2$, the commutative law holds in the set S with respect to $\oplus$.

c. Since $(2 \oplus 3) \oplus 4 = 4$, and $2 \oplus (3 \oplus 4) = 4$, the associative law holds in the set S with respect to $\oplus$.

d. There is a special element 0, which when combined with any number by $\oplus$, does not change the number; i.e., there is an identity element:

$$0 \oplus 0 = 0 \qquad 1 \oplus 0 = 0 \oplus 1 = 1 \qquad 2 \oplus 0 = 0 \oplus 2 = 2$$
$$3 \oplus 0 = 0 \oplus 3 = 3 \qquad 4 \oplus 0 = 0 \oplus 4 = 4$$

e. In the table, the number positioned to the left of the top row or at the top of the left column is the identity element.

Identity Element →

$\oplus$	0	1	2	3	4
0	0	1	2	3	4
1	1	2	3	4	0

f. Each of the elements has an inverse:

$$0 \oplus 0 = 0 \qquad 1 \oplus 4 = 4 \oplus 1 = 0 \qquad 2 \oplus 3 = 3 \oplus 2 = 0$$

g. The inverse of each element (except 0) can be found by subtracting it from 5; e.g., the inverse of 1 is $5 - 1 = 4$. The inverse of the element 0 is 0.

h. The inverse of each element can be found using the table. To find the inverse of 3, find where 0 in its row occurs and read the number at the top of the column, that is, 2.

⊕	0	1	②	3	4
0	0	1	2	3	4
1	1	2	3	4	0
2	2	3	4	0	1
③	3	4	⓪	1	2
4	4	0	1	2	3

Discussion of the responses

1. *About the operation*

The operation table plays an important role in discovering the properties of the operation on the number set S. The properties can be deduced efficiently by making the operation table. For example, the examples of expected responses include the following properties:

        1d expresses the commutative law.

        2a expresses the property of closure.

        2e indicates how to find the identity element.

        2h indicates how to find the inverse of each element.

2. It is important to consider not only what properties the operation $\oplus$ has on the set S = { 0, 1, 2, 3, 4 } but also what features this set has in common with other sets when operation $\oplus$ is performed on set S. Consider, for example, the set Q of rational numbers. (See figure 5.5).

Set S of numbers on the disks under $\oplus$	Set Q of rational numbers under +
Set S is closed with respect to the operation $\oplus$.	Set Q is closed with respect to the operation +.
The commutative law holds in the set S with respect to $\oplus$.	The commutative law holds in the set Q with respect to +.
The associative law holds in the set S with respect to $\oplus$.	The associative law holds in the set Q with respect to +.
The identity element exists in the set S with respect to $\oplus$. $$2 \oplus 0 = 0 \oplus 2 = 2$$	The identity element exists in the set Q with respect to +. $$5 + 0 = 0 + 5 = 5$$
The inverse of each element exists in the set S with respect to $\oplus$. $$0 \oplus 0 = 0$$ $$1 \oplus 4 = 4 \oplus 1 = 0$$ $$2 \oplus 3 = 3 \oplus 2 = 0$$	The inverse of each element exists in the set Q with respect to +. $$0 + 0 = 0$$ $$6 + (-6) = (-6) + 6 = 0$$ $$\frac{1}{3} + \left(-\frac{1}{3}\right) = \left(-\frac{1}{3}\right) + \frac{1}{3} = 0$$

Fig. 5.5

From the observations above, we can see that the set S = {0, 1, 2, 3, 4} has the same properties with respect to the operation $\oplus$ as does the infinite set Q of rational numbers with respect to addition. Thus, we can see that two very different looking sets may have the same structure.

A PROBLEM ABOUT PARALLEL LINES

OSAMU MATSUI

The Lower Secondary School at Chiba University

The Problem and Its Context

The problem

There are parallel lines l and m. Drawing other lines that intersect these lines produces many figures. Find as many properties of the figures as possible that hold wherever the lines that intersect the parallel lines are moved.

Pedagogical context

In teaching geometry, we find that many students cannot think in if-then propositions. The reason is probably that, when considering geometrical figures, students' *free conceptions* —their own ideas—are often hindered by the the focus that is placed on one fixed approach and conclusion. This free conception has an important role to play when students think intuitively about geometrical figures, trying to find a clue to the proof or something new to them. This problem was introduced into classroom teaching as a way to develop students' capacity for free conception.

There are several reasons for considering parallel lines as a topic for this purpose: students are familiar with them from their study in elementary school, the topic is easy to teach, there is good potential for further development, and it is easy to make teaching aids.

The experimental lesson was carried out in group work, but the following content may be used for whole-class instruction with the same objectives.

1. First, students make a new line to intersect two parallel lines l and m using a bamboo stick. Then they rotate this new line around an arbitrary fixed point O, and consider the geometrical figures thus produced.

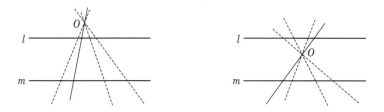

2. Students consider the relation between the position of the fixed point O and the properties of the geometrical figures.
3. From the results of 1 and 2 above, students summarize the basic properties of parallel lines. These properties are illustrated in figure 5.6.

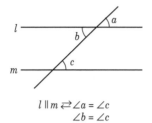

$l \parallel m \rightleftarrows \angle a = \angle c$
$\angle b = \angle c$

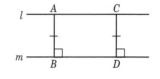

$l \parallel m \rightleftarrows AB = CD$

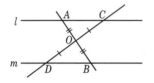

$AO = BO,\ CO = DO$

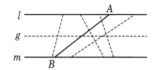

The set of midpoints of AB
is a line, g, parallel to l and m.

Fig 5.6
Properties of Parallel Lines

1. *How to describe*

When students describe the properties they find by adding a new line that intersects the parallel lines, they may be allowed a free style of description to express their observations. However, we supplemented their descriptions with symbolic expressions and figures in order to clarify their conclusions and assumptions. When using symbolic expressions, students are encouraged to write the assumptions and the conclusion separately in sentences and algebraic expressions. When using figures, they are encouraged to draw assumptions in blue and the conclusion in red.

2. *Classroom organization*

The main focus in the lesson is on individual responses. Activities in groups of four to five students can help confirm their ideas about how to view and describe geometrical figures and may also give them new perspectives in viewing the figures. However, if only group activities are used, some students may not develop or express their own opinions. Therefore, we decided to seek and examine the responses of individual students as much as possible by combining group and individual activities.

Expected Responses and Discussion of Them

Examples of expected responses

1. Properties of rectangles and parallelograms
   - lengths of sides, measures of angles
   - conditions for being rectangles and parallelograms

2. If AB and CD are lines intersecting parallel lines l and m, and M and N are midpoints of AB and CD, respectively, then $MN \parallel AC$ and $MN = 1/2\,(AC + BD)$.

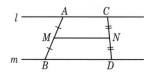

3. In a triangle ABC, if M and N are midpoints of AB and AC, respectively, then $MN \parallel BC$ and $MN = 1/2\,BC$.

4. When the transversals intersect at point O between the parallel lines, students will find the following properties:

- ratio of length of sides
- ratio of areas
- shape of figures (similarity)
- conditions for triangles to be similar

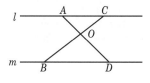

Discussion of the responses

By examining students' worksheets, we can sometimes trace their order of thought. Figure 5.7 is an example.

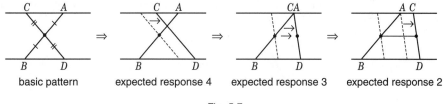

| basic pattern | expected response 4 | expected response 3 | expected response 2 |

Fig. 5.7

We can consider students' responses from the following points of view:

1. How many properties did they describe?
2. Among the properties, how many were correct?
3. How many different properties did they describe?
4. Is there a developmental order among the properties they found?
5. Can the described properties be generalized?

In the case of number 5, we appreciate students conjecturing a general proposition even when they cannot prove it. The ability and attitude to make such a conjecture is necessary for further learning.

As students draw new figures and conjecture their properties, the learning should focus on what students notice and how they develop their observations. Although we used parallel lines as a topic, a similar treatment is possible for a triangle or a circle.

BISECTORS OF BASE ANGLES OF
AN ISOSCELES TRIANGLE

YASUSHI AOYAMA
Shimin Lower Secondary School, Fukui City

The Problem and Its Context

The problem

In the figure below, *BF* and *CD* are the bisectors of the base angles of an isosceles triangle *ABC* (*AB* = *AC*). *CD* intersects *AB* at point *D*; *BF* intersects *AC* at point *E* and the bisector of the exterior angle *ACH* at point *F*.

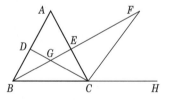

Study this figure from various viewpoints and find as many relations as you can. Write down the relations you find in an orderly way.

Pedagogical context

This problem was posed as the first lesson in eighth-grade geometry. An introduction to logical proofs is one of the major tasks. This problem was used to review what students in the seventh grade had learned about the properties of an isosceles triangle and to help them understand the significance and method of proof by examining the validity of relations they had conjectured. The teacher also intended to foster the abilities to generalize and abstract by having students reason analogically, changing some of the given conditions, or generalizing what they discovered. The lesson can be developed further by giving a logical explanation to the relationships among what the students discovered. This approach will help change their learning attitude from a merely passive or receiving one to an active one.

Expected Responses and Discussion of Them

Examples of expected responses

Viewpoints	Expected responses
Side	1. $AD = AE$, $BD = CE$
	2. $BE = CD$
	3. $BG = CG$, $DG = EG$
Angle	
Vertically opposite angle	4. $\angle DGB = \angle EGC$, $\angle DGE = \angle BGC$
	5. $\angle AEG = \angle FEC$, $\angle AEF = \angle GEC$
Base angle	6. $\angle ABE = \angle CBE = \angle ACD = \angle BCD$
	7. $\angle ABC = \angle ACB$
Other angles	8. $\angle ADG = \angle AEG$, $\angle BDG = \angle CEG$
	9. $\angle BGC = \angle ACH$
	10. $\angle F = 1/2 \angle A$
	11. $\angle BGC = 90° + 1/2 \angle A$
	12. $\angle FCG = 90°$
Shapes and sizes	
Area	13. $\triangle ABE = \triangle ACD$
	14. $\triangle BDC = \triangle CEB$
	15. $\triangle BDG = \triangle CEG$
Shape	16. $\triangle GBC$ is an isosceles triangle.
	17. $\triangle ADE$ is an isosceles triangle.
	18. $\triangle FBC$ is an oblique triangle.
	19. $\triangle GBC$ is an oblique triangle.
	20. $\triangle FGC$ is a right triangle.
Congruence	21. $\triangle ABE \cong \triangle ACD$
	22. $\triangle BDC \cong \triangle CEB$
	23. $\triangle BDG \cong \triangle CEG$
Similarity	24. $\triangle DBC \sim \triangle DGB$
	25. $\triangle ECB \sim \triangle EGC$
	26. $\triangle DEG \sim \triangle CBG$
Others	27. $DE \parallel BC$
	28. The bisector of $\angle A$ passes through the point G.
	29. The diagonals of the quadrangle $ADGE$ are perpendicular to each other.

Discussion of the responses

The responses of a student may be evaluated from two viewpoints: a quantitative one concerned with the number of viewpoints from which the student responded, and a qualitative one concerned with the quality of the viewpoints from which the student responded. For the qualitative viewpoint, weights may be assigned by considering the answers to the following two questions:

1. Is the proposed relation specific or general?
2. Has the proposed relation been studied earlier, or is it a new one?

Figure 5.8 illustrates a way to weight the examples of expected responses listed above.

Viewpoints	Weight		
	1	2	3
Side	1 2 3		
Angle	4 5		
	6 7		
		8 12	9 10 11
Shapes and sizes	13 14 15		
	16 17 18 19	20	
	21 22	23	
		24 25 26	
Others		27	28 29

Fig. 5.8

In the process of examining Response 5 in the list of expected responses above, the teacher can help students grasp the meaning of proposition and show that a proposition can be analyzed into its assumption(s) and its conclusion. Furthermore, the teacher can help students see that properties, such as those of parallel lines and congruent triangles, are a major basis for reasoning.

The teacher can treat Responses 21–23 ($\triangle ABE \cong \triangle ACD$, $\triangle BDC \cong \triangle CEB$, and $\triangle BDG \cong \triangle CEG$) as exercises to acquaint students with the method and description of proof. Responses 1, 2, 3, 8, and 13–17 can be used to provide more insight and understanding of proof.

Although students may know that two triangles are congruent provided that at least three pairs of corresponding sides or angles are equal, there are altogether twenty combinations of pairs of the corresponding sides and angles. Students can examine all twenty combinations. They can also examine the conditions for the congruence of right triangles.

As a further development, the teacher can change the condition "bisectors of base angles" to "medians" or "perpendiculars," and have students determine whether they also are equal to each other.

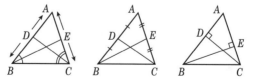

It may be possible to generalize the conditions for an isosceles triangle to a scalene triangle and have students determine whether the equality of lengths, angles, or areas that hold in an isosceles triangle also hold in a scalene triangle.

In examining Response 28 (that the bisector of A passes through the point G), the teacher can help students notice and prove that the point G is at an equal distance from the three sides and that the same relation holds for any triangle. In this way the teacher can work toward establishing the existence of the incenter of a triangle. Students can then consider whether other sets of three lines meet at a point. The teacher can help them determine that three other sets of lines—the three perpendicular bisectors of the sides, the three perpendiculars from the vertices (altitudes), and the three medians—meet at a point. Thus, as students discover the circumcenter, orthocenter, and center of gravity (centroid), the attitude or habit of searching for a new problem (problem formulation and generalization) or properties is fostered.

When a problem situation allows for a variety of conjectures that have mathematical quality, students can learn many things in the process of examining some or all of the conjectures.

A Similar Problem

Midpoints of the Sides of a Quadrilateral

In the figure below, M and N are the midpoints of diagonals AC and BD, respectively, of the quadrilateral $ABCD$. P, Q, R, and S are the midpoints of the four sides AB, BC, CD, and DA, respectively. The four points P, N, R, and M are joined by lines as are S, N, Q, and M. Find as many relations as possible among the sides, angles, and areas that hold in this figure.

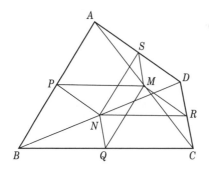

RANKING CLASSES IN A MARATHON RACE

HIROAKI FUKUJU

Kamisugeta Lower Secondary School, Yokohama City

The Problem and Its Context

The problem

The students in four classes A, B, C and D (each with 20 students) participated in a marathon race; 6 students could not run and only observed the race. Table 5.8 shows how each runner ranked within his or her class as well as among all the runners. For example, the first-place runner in class A was fourth among all the runners; the first-place runner in class B ranked sixth overall.

Within-Class Ranking	Classes				Within-Class Ranking	Classes			
	A	B	C	D		A	B	C	D
1	4	6	1	2	11	43	40	44	46
2	9	7	3	5	12	45	42	47	51
3	11	10	14	8	13	49	48	50	55
4	12	13	18	15	14	54	52	56	27
5	20	16	19	17	15	61	53	60	58
6	21	22	23	31	16	65	62	63	59
7	25	24	28	33	17	69	66	64	67
8	26	27	30	36	18	70	72	*	68
9	29	34	32	37	19	71	*	*	73
10	35	39	41	38	20	*	*	*	74

Table 5.8
Individual Ranking by Classes

* Indicates students who observed the race.

Devise as many ways as you can to determine the ranking of classes when the race is an inter-class competition. Write down your methods of ranking the classes, along with the letters. For each ranking method you devise, write down the first-, second-, and third-place classes.

Pedagogical context

The objective is to help students understand central values of group data and consider ways to determine the ranking from various points of view. The problem is concerned with expressing the central tendency of a population numerically, and it can be used at any grade level. Students are likely to take an active interest in tackling the problem since it is real to them.

Expected Responses and Discussion of Them

Examples of expected responses

1. Intuitive ways to decide the ranking in a short time are as follows:
   - *Determine the ranking only by the fastest runner in each class.*
     From row 1 in the table, we see that the first to cross the finish line in class A is runner 4; in B, runner 6; in C, runner 1; and in D, runner 2. Thus, the order of the class ranking is C, D, A, B.
   - *Determine the ranking only by the slowest runner in each class.*
     Because the numbers of runners differ among classes, we look for the last row that has four entries. This is row 17, from which we get the ranking C, B, D, A.
   - *Determine the ranking by comparing the medians of each class.*
     This approach compares performance of an average runner in each class. From row 9, which is in the middle of the 17 rows having full entries, we get the order A, C, B, D.

2. The ways to determine the ranking by considering all participants in each class are as follows:
   - *Rank the classes by the increasing order of their sums or the averages of individual rankings.*
     This approach is more favorable to classes having the fewest runners.
   - *Rank the classes simply by determining the actual number of runners in each.*
     The numbers of runners are 19 for class A, 18 for B, 17 for C, and 20 for D. The class ranking is D, A, B, C.
   - *Regard the individual rankings for all observers as 75 (the last runner was 74), and then rank the classes according to the sums or averages of individual rankings.*
     The sums (averages) are 794 (39.7) for A, 783 (39.2) for B, 818 (40.9) for C, and 800 (40.0) for D. The order is B, A, D, C.
   - *Assign a weighted score to each ranking, and decide the ranking by the sums or averages of members' scores.*
     If all differences of scores between successive rankings are equal, the result would be almost the same as in the previous ranking. Here the differences must be unequal, and giving a negative score to observers would be one approach.

3. The ways to determine the ranking by using samples are as follows:
   - *Determine the ranking by the sums or averages of the top ten runners in each class.*
     Here the sums (averages) of ranking in rows 1 through 10 are 192 (19.2) for A, 198 (19.8) for B, 209 (20.9) for C, and 222 (22.2) for D. The ranking is A, B, D, C.
   - *Regard runners 1–30 as prize winners and determine the class ranking by the number of prize winners in each class.*
     The numbers are 9 for A, 8 for B and C, and 5 for D. So, A is first, B and C tie for second, and D is fourth.
   - *Draw a sample from each class in various ways, give scores to the individual rankings, and decide the class ranking by the sums or averages of the scores of each sample.*

4. Others:

- *Determine the ranking by the number of the fastest runners in each row from 1 to 20.*

  In table 5.9 below, the fastest runner in each row is circled (○). The total numbers of marks are 6 for class A, 7 for B, 3 for C, and 4 for D. The ranking is B, A, D, C.

	A	B	C	D		A	B	C	D
1	4	6	①	2	11	43	㊵	44	46
2	9	7	③	5	12	45	㊷	47	51
3	11	10	14	⑧	13	49	㊽	50	55
4	⑫	13	18	15	14	54	㊾	56	57
5	20	⑯	19	17	15	61	㉝	60	58
6	㉑	22	23	31	16	65	62	63	㊾
7	25	㉔	28	33	17	69	66	㊽	67
8	㉖	27	30	36	18	70	72	*	㊽
9	㉙	34	32	37	19	㉟	*	*	73
10	㉟	39	41	38	20	*	*	*	㉤
					Totals:	6 (A)	7 (B)	3 (C)	4 (D)

* Indicates observers

Table 5.9
The Fastest Runner in Each Row

- *Assign a ranking of 1–4 to the entries in each row, with the lowest ranking to observers, and determine the class ranking by the sums of the row ranking.*

  The sums are 46 for class A, 40 for B, 56 for C, and 54 for D. The class ranking is B, A, D, C.

Discussion of the responses

1. To generate a variety of responses, it is better to give students only two minutes to determine which class will win. This encourages them to intuitively consider the ways to rank the classes.

2. Students may use calculators for the sake of saving time.

3. After students propose various ways to determine the class rankings, a discussion should follow to consider which ways are the most suitable given the aim of the marathon race and the magnitude of the data to be used.

4. Students must naturally realize that in order to be fair in competition, they must agree on the way to determine the ranking, since the result is likely to vary according to the approach adopted.

5. The teacher should point out that this kind of table is easy to use in determining the class rankings, and then direct the lesson toward ways to summarize the data.

Chapter 6

Examples of Teaching in
the Upper Secondary Schools

I N THIS CHAPTER, the main lessons that were taught in the upper secondary schools are presented. Because of space limitations, the records of the teaching are included only for the typical cases. Similar problems, when possible, are added for the readers' information.

THE FORMAT OF EACH SECTION

The Problem and Its Context

1. *The problem:* The problem is stated in the way it was presented to the students.
2. *Pedagogical context:* The purpose of the problem is given along with its connections to the content in the textbook and to the mathematics program.

Expected Responses and Discussion of Them

3. *Examples of expected responses:* Examples of students' expected responses to the problem are classified by viewpoints, including some high-quality responses that some students may make.
4. *Discussion of the responses:* The classification of students' responses, the mathematical values of their responses, how to evaluate the responses, and further mathematical development are detailed.

Record of the Classroom Teaching

5. *Teaching the lesson*: The place of the problem in the whole teaching plan, major questions, and related learning activities are discussed. (Minute details are omitted.)
6. *Remarks after the lesson*: Reflection on the lesson, the time needed, classroom discussion, collecting of students' responses, and further development of the problem are presented.

PROPERTIES OF THE MULTIPLICATION TABLE

YUKIO YOSHIKAWA
The Upper Secondary School at the University of Tokyo

The Problem and Its Context

The problem

The number table in figure 6.1 was produced in accordance with a certain rule. Find as many relationships as possible among the numbers in the table by studying the arrangement of numbers.

1	2	3	4	5	6	7	8	9	10
2	4	6	8	10	12	14	16	18	20
3	6	9	12	15	18	21	24	27	30
4	8	12	16	20	24	28	32	36	40
5	10	15	20	25	30	35	40	45	50
6	12	18	24	30	36	42	48	54	60
7	14	21	28	35	42	49	56	63	70
8	16	24	32	40	48	56	64	72	80
9	18	27	36	45	54	63	72	81	90
10	20	30	40	50	60	70	80	90	100

Fig. 6.1

Pedagogical context

The lesson has the following purposes:

1. Students will learn to consider the relation between a general case and a special example of it, as well as to learn to appropriately use letters as variables in explaining general relations.

2. Students will have opportunities to review what they learned in lessons on "number sequence," such as how to find the sum of an arithmetic progression, how to use the symbol Σ, how to apply the sums $\Sigma\, k$, $\Sigma\, k^2$, and so on.

Therefore, it is appropriate to schedule this lesson after teaching the topic "number sequence." Of course, it would also be worthwhile to teach the lesson independently.

Expected Responses and Discussion of Them

Examples of expected responses

A. *About the arrangements of numbers*

A1. Numbers in each column and row are multiples of 1, 2, 3, …, 10, respectively.

A2. In each column, the difference between each number and the next one is a constant (fig. 6.2); i.e., the numbers in each column constitute an arithmetic progression (AP). The same is true for each row.

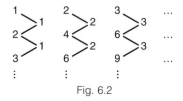

Fig. 6.2

A3. All numbers on the main diagonal from the upper-left to the lower-right are square numbers; i.e., 1, 4, 9, 16, ..., 100.

A4. The numbers are symmetrically arranged with respect to the main diagonal (fig. 6.3).

Fig. 6.3

A5. Figure 6.4 shows a table of differences between the terms along the diagonals from upper right to lower left.

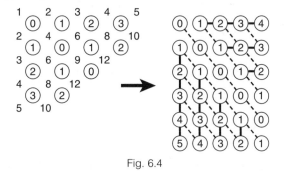

Fig. 6.4

This table has the following properties:

• The terms of the sequences along the dotted lines are the same number.

• The terms of the sequences along the solid lines are the terms of the sequence of natural numbers.

• When the differences are taken always from upper right to lower left and expressed by signed numbers, the terms on opposite sides of the 0 - 0 - 0 - ... diagonal have opposite signs (fig. 6.5).

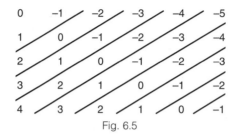

Fig. 6.5

- The terms along the diagonals from upper right to lower left are APs.

A6. Figure 6.6 shows a table of differences in which the differences are taken between terms along the diagonals from upper left to lower right. This table has properties analogous to those of the table in figure 6.4. For example, whereas in figure 6.4 the same number occurs on the diagonals from upper left to lower right, in figure 6.6 the same number occurs on the diagonals from upper right to lower left.

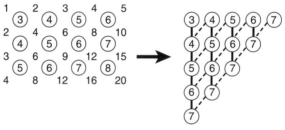

Fig. 6.6

A7. Figure 6.7 shows a procedure to generate the terms in the sequence 1, 4, 12, 32, 80, ...

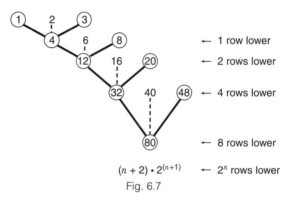

$(n + 2) \cdot 2^{(n+1)}$ ← 2^n rows lower

Fig. 6.7

To find the terms 1, 4, 12, 32, 80, ... , first consider 1 and 3 in the first row, skipping the term between them (2). Their sum, 4, is one row below the skipped term. Next consider 4 and 8 in the second row again skipping the term between them. Their sum, 12, is two rows below the skipped terms. Similarly, the sum of 12 and 20 in the fourth row is four rows below the skipped term.

The sum of 32 and 48 in the eighth row is eight rows below the skipped term. In the $(n + 1)$th step, the sum will be $(n + 2) \cdot 2^{n+1}$ and in the 2^n rows lower (from the previous row).

Analogous patterns can be found by skipping any other odd number of terms.

B. *About the sums of the numbers*

B1. The sum of all the numbers in each row or column is a multiple of 55.

B2. Using row 1 in the table as the column numbers and column 1 as the row numbers, you can compute the sum of the numbers enclosed in a rectangle (fig. 6.8) as follows: (a) add the numbers of the columns that are included in the rectangle; (b) add the numbers of the rows that are included; and (c) multiply the two sums. Thus, the sum of the numbers enclosed in the rectangle in figure 6.8 can be found by computing $(5 + 6 + 7 + 8 + 9) \times (6 + 7 + 8)$.

1	2	3	4	5	6	7	8	9	10
2	4	6	8	10	12	14	16	18	20
3	6	9	12	15	18	21	24	27	30
4	8	12	16	20	24	28	32	36	40
5	10	15	20	25	30	35	40	45	50
6	12	18	24	30	36	42	48	54	60
7	14	21	28	35	42	49	56	63	70
8	16	24	32	40	48	56	64	72	80
9	18	27	36	45	54	63	72	81	90
10	20	30	40	50	60	70	80	90	100

Fig. 6.8

B3. When a number in the table is taken as a pivot, the sum of the two numbers in a row or column located symmetrically about the pivot is two times the pivot number. This property is illustrated on the left side of figure 6.9. The right side of figure 6.9 shows a further development of the property.

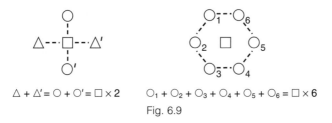

$\triangle + \triangle' = \bigcirc + \bigcirc' = \square \times 2$ $\bigcirc_1 + \bigcirc_2 + \bigcirc_3 + \bigcirc_4 + \bigcirc_5 + \bigcirc_6 = \square \times 6$

Fig. 6.9

B4. The difference between the sums of the numbers on opposite corners of any square of size 1×1 is 1 (fig. 6.10).

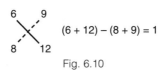

$(6 + 12) - (8 + 9) = 1$

Fig. 6.10

In general, for any $m \times n$ rectangle of numbers, as in figure 6.11, $\Box + \Box' - (\triangle + \triangle')$ = mn. Thus, the difference mn is constant for any rectangle of numbers of size $m \times n$.

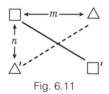

Fig. 6.11

B5. Ratio of the sums V, I, P in a rectangle. V is the sum of the numbers at the vertices of the rectangle. I is the sum of the numbers inside the rectangle (i.e., not part of any side). P is the sum of the numbers on the perimeter of the rectangle (including the vertices). The ratio $V : I : P$ is constant for any 3×5 rectangle (fig. 6.12).

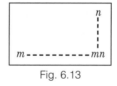

Fig. 6.12

Generally speaking, for rectangles of dimensions $m \times n$,

$$V : I : P = 4 : (m - 2)(n - 2) : 2(m + n - 2).$$

The ratio is constant so long as the dimensions of the rectangle are not changed.

C. *About the products of the numbers*

C1. The number in the mth row and nth column is mn.

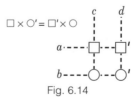

Fig. 6.13

C2. For any rectangle formed by rows and columns, the product of the two numbers at the ends of a diagonal is equal to that of the numbers at the ends of the other diagonal (fig. 6.14).

$$\Box \times \bigcirc' = \Box' \times \bigcirc$$

Fig. 6.14

C3. If the rectangle in C2 is a square, the products of all the numbers on a diagonal is equal to that of all the numbers on the other diagonal.

D. *About number sequences*

In these responses, a represents the sum of a specific set of numbers obtained in the ith step.

D1. Figure 6.15 shows number sequences created by perpendicular lines along parts of like-numbered rows and columns.

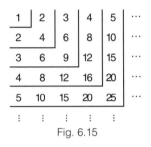

Fig. 6.15

Where the subscript refers to the row number,

$a_1 = 1$ $\qquad\qquad\qquad = 1$

$a_2 = 2 + 4 + 2$ $\qquad\qquad = 8$

$a_3 = 3 + 6 + 9 + 6 + 3$ $\qquad = 27$

$\vdots$

$a_n = n\,(1 + 2 + \cdots + n) + (1 + 2 + \cdots + (n - 1))\,n = n^3$

D2. In figure 6.16, a 2×2 square is moved down one row and one column to the right.

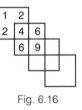

Fig. 6.16

Where the subscript refers to the row in which the top of the square is located,

$a_1 = 1 + 2 + 2 + 4$ $\qquad\qquad = 9$

$a_2 = 4 + 6 + 6 + 9$ $\qquad\qquad = 25$

$\vdots$

$a_n = n^2 + 2n\,(n + 1) + (n + 1)^2 \qquad = (2n + 1)^2$

The investigation of this property can be developed further by enlarging the square. For a 3×3 square,

$a_1 = 36$

$a_2 = 81$

$\vdots$

$a_n = (3n + 3)^2$

Generally, for a $k \times k$ square, $a_n = (kn + k(k - 1)/2)^2$

Note: The case of the slope being (–1), which is developed in response D6(d), is a special case of this general one, where $k = 1$.

A further development is to change the direction in which the square is moved. For example, when the square is moved directly down one row (instead of down and to the right), we have $a_1 = 9$, $a_2 = 15$, $a_3 = 21$, ... , $a_n = 6n + 3$.

When the square of $k \times k$ is moved similarly, we have

$$a_n = 1/2 \times k(k + 1)(kn + k(k - 1)/2).$$

Still further developments are possible; for example, by using different slopes for the movement or by changing the square to a rectangle.

D3. Figure 6.17 shows a sequence of cross shapes within the number table.

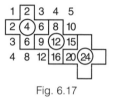

Fig. 6.17

Where the subscript refers to the row in which the top of the cross is located,

$a_1 = 4 + 2 + 2 + 6 + 6 = 20$

$a_2 = 12 + 8 + 9 + 16 + 15 = 60$

$\vdots$

$a_n = 10n(n + 1)$

Note: In this case, the sequence 4, 12, 24, ... is obtained by taking slope –2 movement in the response D6, and for the second part the sum of a pair of terms symmetrically placed with respect to the pivot (marked by ○) is twice the pivot as in response B3.

D4. In figure 6.18, a 1×2 rectangle is moved one row down and one column to the right.

Fig. 6.18

Where the subscript refers to the row in which the rectangle is located,

$a_1 = 1 + 2 = 3$

$a_2 = 4 + 6 = 10$

$\vdots$

$a_n = 2n^2 + n$

The investigation can be developed further by moving the rectangles in a different direction or by taking another rectangle to make pairs (fig. 6.19).

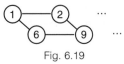

Fig. 6.19

Another further development is to increase the number of numbers in a rectangle (fig. 6.20).

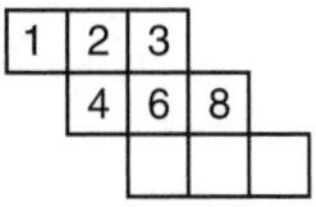

Fig. 6.20

Note: This also may be regarded as a further development of D2 to the case of a rectangle.

D5. Figure 6.21 shows a series of squares within squares.

9	12	15	18	21	24
12	16	20	24	28	32
15	20	25	30	35	40
18	24	30	36	42	48
21	28	35	42	49	56
24	32	40	48	56	64

Fig. 6.21

Where the sequence of subscripts begins with the innermost square, compute the sum of the terms on the perimeter of the square as follows,

$a_0 = 121$ $25 + 2 \times 30 + 36$

$a_1 = 363$ $16 + 2 \times 20 + 2 \times 24 + \cdots + 2 \times 42 + 49$

$a_2 = 605$ $9 + 2 \times 12 + 2 \times 15 + \cdots + 2 \times 56 + 64$

$\vdots$

$a_n = 121 \times (2n + 1)$

A further development is to change the location of the a_0 (innermost) square:

$a_n = $ (sum of terms in the first square) $\times (2n + 1)$

Another further development is to make a_0 a 1×1 square (fig. 6.22).

6	8	10	12	14
9	12	15	18	21
12	16	20	24	28
15	20	25	30	35
18	24	30	36	42

Fig. 6.22

In this case, compute the sum of the terms on the perimeter of square as follows, where the underlined numbers represent the number of terms on the perimeter.

$a_1 = 160 = 20 \times \underline{8}$

$a_2 = 320 = 20 \times \underline{16}$

$\vdots$

$a_n = 20 \times \underline{8n}$

The example above can be used for the original case (fig. 6.21) by regarding the underlined numbers as the number of terms as in figure 6.23.

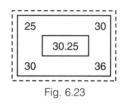

Fig. 6.23

$a_0 = 30.25 \times \underline{4} = ((25 + 2 \times 30 + 36) \div 4) \times 4$

$a_1 = 30.25 \times \underline{12}$

$\vdots$

$a_n = 30.25 \times (\underline{8n + 4})$

D6. A sequence of sums can be formed by using the terms shown on the diagonals in figure 6.24. Regard the table in figure 6.24 as being extended beyond the tenth column and row.

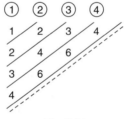

Fig. 6.24

(a) The case of slope 1:

a_1 $= 1$

$a_2 = 2 + 2$ $= 4$

$a_3 = 3 + 4 + 3$ $= 10$

$a_4 = 4 + 6 + 6 + 4$ $= 20$

$\vdots$

$a_n = 1 \times n + 2 \times (n - 1) + \cdots + 2 \times (n - 1) + 1 \times n = 1/6 \times n(n + 1)(n + 2)$

One way to find the sum for a_n is by applying $\Sigma\, k^2$ and $\Sigma\, k$:

  1   4   10   20   25 ...

      3   6   10   15   ... (triangular numbers)

$a_n = \Sigma\, k\,(k + 1)/2 = 1/2\, \Sigma\, (k^2 + k)$

The sum can also be derived from the structure of the table:

$$1 \times 1 \qquad\qquad 1 \times 2 \qquad\qquad 1 \times 3$$
$$2 \times 1 \qquad\qquad 2 \times 2 \qquad\qquad 2 \times 3$$
$$3 \times 1 \qquad\qquad 3 \times 2 \qquad\qquad 3 \times 3$$
$$\Rightarrow \sum i(n - i + 1) = (n + 1) \sum i - \sum i^2$$

(b) The case of slope 2 (fig. 6.25):

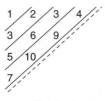

Fig. 6.25

$$a_1 = 1 \times 1 \qquad\qquad = 1$$
$$a_2 = 1 \times 2 + 3 \times 1 \qquad = (1 + 3) + 1 = 5$$
$$a_3 = 1 \times 3 + 3 \times 2 + 5 \times 1 = (1 + 3 + 5) + (1 + 3) + 1$$
$$= 9 + 4 + 1 = 14 \text{ (sum of square numbers)}$$
$$\vdots$$
$$a_n = n(n + 1)(2n + 1)/6$$

(c) Reference to cases where the slope is greater than 2 are omitted here for brevity, though such developments are possible.

(d) Cases of negative slope are also possible. The case of slope (−1) is examined below. A consideration similar to that for a positive slope is possible for the sum of the terms up to the kth.

$$a_1 = 1 + 4 + 9 + \cdots + k^2 = k(k + 1)(2k + 1)/6$$
$$a_2 = 2 + 6 + 12 + \cdots + k(k + 1) = k(k + 1)(k + 2)/3$$
$$\vdots$$
$$a_n = \sum i(i + n - 1) = k(k + 1)(2k + 3n - 2)/6$$

Discussion of the responses

1. *Relation to the structure of the table*

In considering any property of the table, one must proceed by relating it to the basic structure of the table. For example, in the case of D6 (d), ② − ① equals

$$1/3 \times k(k + 1)(k + 2) - 1/6 \times k(k + 1)(2k + 1) = 1/2 \times k(k + 1),$$

and this result is due to the following relation:

$$
\begin{array}{rl}
② & 2 + 6 + 12 + \cdots + k(k + 1) \\
-① & 1 + 4 + 9 + \cdots + k^2 \\
\hline
& 1 + 2 + 3 + \cdots + k = k(k + 1)/2
\end{array}
$$

This relation is confirmed by direct observation, as shown in figure 6.26.

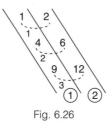

Fig. 6.26

The sums ②, ③, … may also be found from ①, when this fact is noticed.

2. *Features of the table*

The most important features in the structure of the table might be the following:

a) It is a multiplication table. In other words, the number in the mth row and nth column is mn.

b) Each row and each column are arithmetic progressions.

The fact that the sequences are APs is related to most of the properties mentioned in item 1. Among them, the ratio property of V, I, and P mentioned in B5 can be found in other tables, such as those for the addition facts, the subtraction facts, or a 10×10 arrangement of integers from 1 to 100. In all these tables, the ratio is invariably $4 : (m - 2)(n - 2) : 2(m + n - 2)$.

Actually, the fact that the ratio $V : I : P$ is invariate can be seen in a more general table whose rows and columns are APs. For example, the ratio $V : I : P$ is invariate in the table in figure 6.27 whose rows are APs with a common difference 2 and whose columns are also APs with a common difference 3.

3	5	7	9	11	13	…
6	8	10	12	14	16	…
9	11	13	15	17	19	…
12	14	16	18	20	22	…
15	17	19	21	23	25	…
⋮	⋮	⋮	⋮	⋮	⋮	

Fig. 6.27

When each row of a table is an AP with a common difference, if the columns are also APs, then their common differences must be equal.

Moreover, the range of tables that have the property that the ratio $V : I : P$ is equal to that in B5 can be extended further. That is, the same result as that in B5 may be obtained when the condition changes from "common differences of rows are equal" to the condition that "common differences of successive rows make an AP." The former condition is a special case of the latter condition, where the difference is 0.

The equality $V : I : P = 4 : (m - 2)(n - 2) : 2(m + n - 2)$ can be also interpreted as shown in figure 6.28.

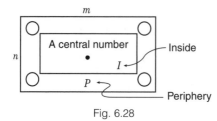

Fig. 6.28

The number of terms in V is 4 (the number of ○'s in the above figure).

The number of terms in I is $(m-2)(n-2)$.

The number of terms in P is $2(m+n-2)$.

The V, I, and P are equal to (a central number) × (the number of terms), respectively.

Record of the Classroom Teaching

Teaching the lesson

The lesson was taught after the teaching of an ordinary unit on number sequence. The lesson was taught in three parts of two class periods each.

Worksheets with the problem printed on them were prepared. The tables were printed on ten large sheets of paper for use during the teacher's explanation. The lesson was taught in a room equipped with special audiovisual aids and a color television system.

The lesson progressed as follows:

1. The teacher presented the problem using the television in the first session. Then the teacher distributed the worksheets. During the next forty-five minutes, individual students presented their work without any specific leads by the teacher.

   Students presented the observations below using the television and partially colored tables:

   • Each row and each column are APs.

   • The numbers in the table are arranged symmetrically about the diagonal.

   • All the numbers on the diagonal are square numbers.

   Comments on these observations were offered by students:

   "What about the sequence 2, 6, 12, 20, ... ?"

   "These sequences are made by taking numbers like moves of the bishop in chess."

   "Now, I will try taking numbers like moves of the knight in chess."

   At this point, the teacher entered the discussion and summarized the students' observations by introducing the concept of slope, and told the students to continue. The teacher refrained from saying too much.

   D1. The sum of numbers in each partition in figure 6.29 equals ○ × □.

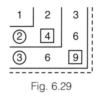

Fig. 6.29

"$2 \times 4 = 2^3$, $3 \times 9 = 3^3$, ... So, each term must be n^3."

"Are you sure it holds always?"

"Yes, I will prove it. The nth term is
$$(n \times 1 + n \times 2 + \cdots + n \times n) \times 2 - n^2$$
$$= n(1 + 2 + \cdots + n) \times 2 - n^2$$
$$= n \times n\,(n + 1) - n^2$$
$$= n^3 + n^2 - n^2$$
$$= n^3\text{''}.$$

"I see."

(Voices of admiration were heard.)

The teacher collected the worksheets that were distributed in the beginning of the lesson, passed out new ones, and asked the students to summarize their own ideas before the next class.

2. In the next part of the lesson the teacher sometimes intervened by asking leading questions or by explaining. Because only a few students were prepared, the teacher first assigned individual work for about thirty-five minutes. A presentation of the results follows:

- An explanation was given of the features of the sequences formed by using various slopes referred to in D6. Since no students referred to the sums of these sequences, the teacher asked them to find the sum of each sequence in D6. Some students found this assignment difficult. About forty more minutes were spent working.

D2.

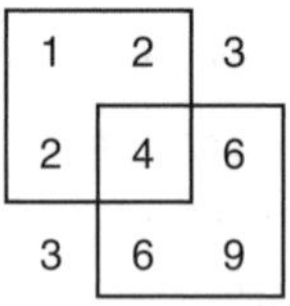

1	2	3
2	4	6
3	6	9

Fig. 6.30

The general term is $(2n + 1)^2$.

(The explanation was given on the chalkboard.)

"The direction in which a square may be moved can be changed."

Actually making the sequences was left to the students to try themselves.

D5.

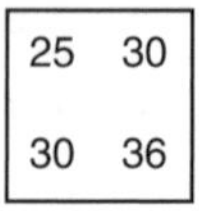

25	30
30	36

Fig. 6.31

For sequences of the type obtained by extending the figure above, the teacher gave a suggestion about an extension from the 1×1 square.

3. For the third part of the lesson, the teacher led the development. The students used calculators. After showing students the table in which rectangles were drawn in color, the teacher assigned an exercise to find the sum for a rectangle. The teacher pointed out that he got answers much more quickly than they did.

Next the teacher explained the meaning of V, I, and P, and asked students to find the ratio $V:I:P$. They studied the ratio for rectangles of the same size ($m \times n$) located in different parts of the table. The students observed that the ratios were all equal and tried to determine why this was the case. The teacher hinted that he had gotten answers much more quickly than the students. About one-third of the students noticed that the answer could be found by (a central number) × (number of terms).

Remarks after the lesson

1. *Examples of responses found in students' worksheets*

Though it was impossible in the actual lesson to consider so many different ideas, the teacher found most of the expected responses by carefully examining students' worksheets. The teacher also found poor expressions or a lack of generalization in the worksheets.

Responses B2 and B5 were not found in the students' worksheets. It is plausible that the ratio $V:I:P$ of B5 was not noticed, but why B2 was not noticed is an open question: Is it not natural to hit on the idea of the sum of the terms enclosed by a rectangle?

2. *The students' attention to sequences along diagonals*

Though the teacher had expected that skillfully leading the students would be necessary in order for them to notice the sequences along diagonals, nearly one-third of them noticed the sequences; and among them, five or six students in each class focused on the sequence with slope (–1) or another slope. The fact that square numbers are lined up on the diagonal that is a line of symmetry seemed to lead the students to focus on these sequences.

3. *Students' impressions*

The number of students who showed good progress was larger than expected, according to results on students' worksheets and the reports they presented afterward. Individual students indicated that (a) they could complete their ideas, which others did not observe; (b) when they thought through one property, it gradually developed into another, and then to an unexpected stage; and (c) they could give the proof of what they found by using letters as variables.

Similar Problems

The Problem of §8 in Chapter 4, "Pascal's Triangle"

A Number Sequence
The table in figure 6.32 is formed by arranging numbers according to a certain rule:

Line	Number of terms									
	1	2	3	4	5	6	7	8	9	10
1	1	2	3	4	5	6	7	8	9	10
2	3	5	7	9	11	13	15	17	19	
3	8	12	16	20	24	28	32	36		
4	20	28	36	44	52	60	68			
–	–	–	–	–	–	–				
–	–	–	–	–	–					
○	□									

Fig. 6.32

(1) The number 112 appears twice in the table. At what positions does it occur? □th term of ○th line and □th term of ○th line
The last line has only one number. What is its line number? What is the last term? ○th line; the last term is □.

(3) There are many rules that can be found in the table other than the original one. List as many rules as possible in an organized format.

Matrix
There is a set of 2×2 matrices $\begin{pmatrix} a & c \\ b & d \end{pmatrix}$ with a condition that $a + b = c + d$ and a, b, c, d are real numbers.
List as many properties of this set as possible.

COMMON PROPERTIES AMONG A GROUP OF GRAPHS

SHŪEI SASAKI
Yamagata Minami Upper Secondary School, Yamagata

The Problem and Its Context

The problems

1. Draw graphs of the following functions defined over the set of real numbers. List as many properties as you can that are common to at least two of them.

   a. $y = -x^3 + 1$ b. $y = x^3 - 3x^2 + 2$

   c. $y = 3x^4 - 4x^3 + 1$ d. $y = 3x^4 - 2x^3 - 3x^2 + 2$

   e. $y = (x^2 - 1)^2$

   Example: The graphs pass through the point (0, 1): a, c, e

2. Draw graphs of the following functions whose domains are $0 \le x \le 1$. List as many properties as you can that are common to at least two of them.

   f. $y = 6x(1 - x)$ g. $y = 12x^2(1 - x)$

   h. $y = 20x^3(1 - x)$ i. $y = 30x^4(1 - x)$

Pedagogical context

The problems may be used as a summary after teaching an ordinary course on differentiation. Such a course includes the meaning of a differential coefficient, derivatives and their computation, and the application of differentiation. The application of differentiation includes the topics of increase and decrease, the maximum and minimum values of a function (derived from the relation between changes in values of a function and the signs of the values of its derivative), and the graphs of the functions.

Usually, such lessons end by having students draw the graphs of given functions, or in a one-by-one study of them. In this lesson, after showing students the graphs of several functions, the teacher asked them to list the properties common to some of the graphs. The aims were to develop students' viewpoints by which to look at graphs, to classify them, and to summarize the general features of graphs of polynomial functions and the relations between the algebraic expressions of such functions and the shapes of their graphs.

Expected Responses and Discussion of Them

Examples of expected responses

Figure 6.33 shows the graphs of the functions in problem 1.

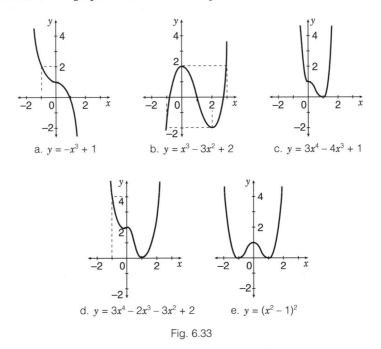

a. $y = -x^3 + 1$ b. $y = x^3 - 3x^2 + 2$ c. $y = 3x^4 - 4x^3 + 1$

d. $y = 3x^4 - 2x^3 - 3x^2 + 2$ e. $y = (x^2 - 1)^2$

Fig. 6.33

Types of responses

1. *Curves*—the graphs are curves (a, b, c, d, e).
2. *Degree of the functions*—they are graphs of a cubic function (of the third degree) (a, b), and so on.
3. *Passing through points*—the graphs pass through the point (1, 0) or (0, 2) (a, b, c, d, e), and so on.
4. *Global maximum and minimum*—the graphs have a global minimum (c, d e), and so on.
5. *Local maximum and minimum*—the graphs have local minimal values at two points (d, e), and so on.
6. *$y' = 0$ but not a maximum or minimum*—the equation $y' = 0$ has a double root (a, c), and so on.
7. *Point symmetry*—the graphs are point-symmetric (a, b), and so on.
8. *Line symmetry*—the graphs are not line-symmetric about the y-axis (a, b, c, d), and so on.
9. *Range of values*—the ranges are $y \geq 0$ (c, d, e), and so on.
10. *Quadrants where the graph exists*—the graphs are only in the first and second quadrants (c, d, e), and so on.

11. *Relation with the x-axis*—the graphs are tangent to the x-axis (c, d, e).

12. *Relation with the y-axis*—the graphs intersect the y-axis (a, b, c, d, e).

13. *Number of points of intersection of the line y = k*—line $y = k$ parallel to the x-axis can be drawn so as to intersect the graph in more than 2 points (b d, e).

14. *Increase, decrease*—the graphs are monotonically increasing on $x \geq 2$ (b, c, d, e), and so on.

15. *Points where y' = 0*—there are two points where $y' = 0$ (b, c).

16. *Value of y for x = a*—$y > 0$ for $x = 2$ (c, d, e).

Figure 6.34 shows the graphs of the functions in problem 2.

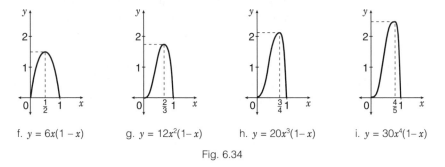

f. $y = 6x(1 - x)$ g. $y = 12x^2(1 - x)$ h. $y = 20x^3(1 - x)$ i. $y = 30x^4(1 - x)$

Fig. 6.34

Types of responses

1. *Curve*—the graphs are curves (f, g, h, i).

2. *Shape of curve*—the graphs have a mountain-like shape (f, g, h, i).

3. *Symmetry*—the graphs are not line-symmetric (g, h, i).

4. *Passing through the point (0, 0) or (1, 0)*—the graphs pass through the origin (f, g, h, i), and so on.

5. *Range of values*—$y \geq 0$ (f, g, h, i).

6. *Increase, decrease*—the graphs decrease after increasing (f, g, h, i).

7. *Local maximum and minimum*—the graphs have only one local maximum (f, g, h, i).

8. *The point where y' = 0*—the graphs have two points where $y' = 0$ (g, h, i).

9. *Tangent to the x-axis at the origin*—$y' = 0$ at $x = 0$ (g, h, i).

10. *Global maximum and minimum*—the graphs have a global maximum (f, g, h, i).

11. *Values of the global maximum*—$1 < y_{max} < 3$ (f, g, h, i).

12. *Interval of values of x that give a maximum*—$1/2 \leq x < 1$ (f, g, h, i).

13. *Value of x that gives a maximum*—the function of nth degree is maximal at $x = (n - 1)/n$ (f, g, h, i).

14. *Number of points of intersection of y = k*—there are two values of x for some value of y (f, g, h, i).

15. *Number of points of intersection of y = x*—the graphs intersect $y = x$ at three points (g, h, i).

16. *Area*—when the area enclosed by the curve and the x-axis is divided into two parts by drawing a line through the maximum point and parallel to the y-axis, the area of the left part is larger than the right part (g, h, i).

Discussion of the responses

Because the problems can be answered intuitively or by just observing the graphs, students can find a variety of properties. It is important for the teacher to lead students to identify and present many properties by using what they have learned, regardless of their skills in expressing the properties.

Next, have students reflect on the viewpoints they used in observing the graphs and then categorize the observations according to the viewpoints (the "types of responses" listed in the previous section), regardless of their differences in expression. Doing this helps the students to understand more clearly the viewpoints from which they look at the graphs. For problem 1, the activities end at this stage.

For problem 2, it is worthwhile to have the students carefully investigate to find out whether any of the viewpoints are mathematically equivalent. For example, consider whether the following responses are different views of the same property.

Shapes of curve—the graphs have a mountain-like shape.

Increase-decrease—the graphs decrease after increasing.

Local maximum and minimum—the graphs have only one local maximum.

Number of points of intersection with $y = k$—there are two values of x for some value of y.

Such an investigation helps to stimulate students to further higher-order considerations.

Possibility of a further development

As mentioned earlier, the problem was tried out in a lesson whose purpose was to summarize students' learning of the topic of differentiation. But if it were used in a lesson after the topics of integration, the limit of values of a function, or the second-order derivatives, other viewpoints or properties could be observed by the students, such as the following:

Area—the area enclosed by the x-axis and the graph is 1 in all four cases of problem 2.

Point of inflection—the graphs in problem 1 may be characterized by the number of inflection points according to whether they graph functions of the third or fourth degrees.

Convexity of graphs—the response, "Number of points of intersection of $y = x$" in problem 2 is related to the convexity of graphs.

The following are other examples of the further development of problem 1:

- The general features of polynomial functions of the nth degree, using the limit of the values of a function.
- The character of the graph at the point where $y' = 0$ when the function has no local maximum or minimum, using the point of inflection.
- The properties of odd and even functions, using the symmetry of the graphs.

Record of the Classroom Teaching

Teaching the lesson

The lesson was taught during two periods after teaching the topic of differentiation and its application and before the topic of integration and its application.

The numbered actions below are followed by the number of minutes (in parentheses) needed for students' activities. The asterisk indicates what was said or done by the teacher.

1. The students drew graphs of the functions in problems 1 and 2. (35 minutes)

   * As you have already learned, it is important to draw graphs. Today, we will once again draw graphs as a summary of our earlier lessons.

   * Because the domain of the functions in 2 is $0 \leq x \leq 1$, you need to draw the graphs only for that domain.

   (Some students were drawing graphs without paying attention to the domain.)

2. The teacher showed the correct graphs and had students correct their errors, if any. (10 minutes)

   * Most of you seem to have drawn the graphs correctly, but we need to be sure of it. If you find any errors, correct them after identifying their source.

3. The teacher distributed the worksheets for problem 1. Using the given example as a guide, students tried to identify common properties. (15 minutes)

   * Look at the five graphs for problem 1. There are many properties that are common to at least two of them, such as the property in the example. Write down as many such properties as you can.

4. Each student presented his or her responses, and the teacher ascertained their validity with the whole class. (10 minutes)

   * I want you to present the responses you wrote down, and also any others you found afterward.

5. The teacher helped the students classify their responses according to the various points of view. (5 minutes)

   * We have so many properties. Are there any among them that are the result of looking at the graphs from the same point of view, even though they may be expressed differently? If so, let's arrange them accordingly.

6. After distributing the worksheets for the graphs in problem 2, the teacher had students respond and submit their results. (10 minutes)

   * Now, study the graphs in problem 2 and write down your findings. Pay attention to the points of view from which you examined the graphs in problem 1.

7. Each student presented the properties he or she had observed, and then the teacher categorized them while ascertaining their validity with the whole class. (10 minutes)

   * I want you to openly present your observations. Please consider whether your responses are different in nature from what your classmates have presented.

8. The teacher showed students the table of "examples of expected responses" and emphasized the multitude of viewpoints from which the graphs can be observed. (5 minutes)

  * We have so many properties. Maybe you have considered even more. When we consider all of them, we get a table like this.

  * Some responses make a reference to area. In each graph, the area enclosed by the x-axis and the graph is 1. This is related to integration, which will be discussed in the next lesson. Thus, you can look at the graphs from a variety of viewpoints based on your own knowledge.

Students' responses

Apart from some differences in aptness of students' expression of properties, most were correct. Only three mistakes were identified. Even low-achieving students and those who had been "negative" in previous learning actively participated in the lesson. The heightened interest was due partly to the intuitive approach used in the lesson, but it was due more to the opportunity given students to use their own ways of thinking and ideas that were based on their experience, and to the fact that their responses were identified as correct and given recognition by all students during the discussion. For low-achieving students who had experienced less success up to this point, this appeared to be a happy learning experience.

On the other hand, some high-achieving students were perplexed that they were asked to answer such simple questions. They thought the answers seemed intuitively self-evident and became quite cautious. Afterward, the high-achieving students realized that the review had helped to deepen their understanding.

Remarks after the lesson

Overall, the lesson progressed according to the following steps:

1. Each student considered common properties among the graphs and individually recorded his or her findings on a worksheet.

2. The students presented their findings to the class. The whole class then discussed and categorized the responses.

The teacher tried to involve as many students as possible in presenting their findings. However, it was impossible to discuss all of the students' work because the time was limited.

The teacher studied the worksheets of all students in the two classes, one with 45 students (A) and the other with 41 students (B), and tabulated the responses. Tables 6.1 and 6.2 list the types and numbers of responses for problems 1 and 2, respectively.

Properties were regarded as the same if they had a common viewpoint, even if expressed differently. The number of responses by individual students ranged from two to thirteen and had no significant correlation with students' usual level of achievement. It was also observed that high-achieving students were likely to give more abstract or general responses.

The fact that the average number of responses for problem 2 is larger than that for problem 1 may be due to the effect of what students learned working on problem 1. For

	Type of Response	Class	
		A	B
1.	Curve	8	17
2.	Degree of the functions	10	10
3.	Passing through the points $(1, 0)$ or $(0, 2)$	37	38
4.	Global maximum and minimum	39	20
5.	Local maximum and minimum	37	31
6.	$y' = 0$, but not a maximum or minimum	4	6
7.	Point symmetry	11	9
8.	Line symmetry	3	5
9.	Range of values	32	21
10.	Quadrants where graph exists	23	30
11.	Relation with the x-axis	10	18
12.	Relation with the y-axis	3	7
13.	Number of points of intersection of the line $y = k$	1	3
14.	Increase-decrease	23	20
15.	Points where $y' = 0$	31	10
16.	Value of y for $x = a$	5	1
Average		6.2	6.0

Table 6.1
Number of Each Type of Response for Problem 1

	Type of Response	Class	
		A	B
1.	Curve	4	5
2.	Shape of curve	23	33
3.	Symmetry	7	8
4.	Passing through the points $(0, 0)$, $(1, 0)$	41	40
5.	Range of values	37	39
6.	Increase-decrease	28	26
7.	Local maximum and minimum	44	39
8.	The point where $y' = 0$	9	7
9.	Tangent to the x-axis at the origin	26	24
10.	Global maximum and minimum	43	39
11.	Value of the global maximum	15	11
12.	Interval of the values of x to give a maximum	7	6
13.	Value of x to give a maximum	4	4
14.	Number of points of intersection of the line $y = k$	30	29
15.	Number of points of intersection of the line $y = x$	2	0
16.	Area	4	2
Average		7.2	7.6

Table 6.2
Number of Each Type of Response for Problem 2

example, there were four "Number of points of intersection of the line $y = k$" responses for problem 1. For problem 2, there were fifty-nine such responses.

The tables of the types of responses show that many student responses were a consequence of using the viewpoints "increase-decrease," "local maximum and minimum," and "global maximum and minimum" learned in the lesson. It seems clear that there are considerable differences in the types of student responses when a lesson such as this is given. The fact that low-achieving students gave unexpected responses not given by the other students may be regarded as one of the important features of the open-ended approach.

NECESSARY CONDITION AND SUFFICIENT CONDITION

SHŪEI SASAKI
Yamagata Minami Upper Secondary School, Yamagata

The Problem and Its Context

The problem

1. At the end of each of the following statements, write "T" if the statement is true, and "F" if it is false.

    A. If $x = 2$, then $x^2 = 4$.

    B. If $x = 2$, then $x^2 - 4x + 4 = 0$.

    C. If $x = 2$, then $x^2 - 2x + 1 = 0$.

    D. If x is a divisor of 3, then x is a divisor of 12.

    E. If x is a divisor of 6, then x is a divisor of 12.

    F. If x is a divisor of 9, then x is a divisor of 12.

    G. If one of the pair (x, y) is positive and the other negative, then xy is negative.

    H. If xy is negative, then one of the pair (x, y) is positive and the other negative.

    I. If both x and y are positive, then xy is positive.

    J. If xy is positive, then both x and y are positive.

2. Fill in the following blank with an expression about x or y that fits it. Write as many expressions as you can.

    If both x and y are even, then _____ is even.

3. Fill in the blank with an expression about x or y that fits it. Write as many expressions as you can.

    A. If either x or y is odd and the other even, then _____ is odd.

    B. If _____ is odd, then either x or y is odd and the other even.

    C. If both x and y are odd, then _____ is odd.

    D. If _____ is odd, then both x and y are odd.

Pedagogical context

The purpose of these problems is to help students learn about hypothetical propositions, how to determine their truth, and the diversity of possible antecedents and consequents. On the basis of students' previous learning, the teacher wants to help them clearly understand the definitions of *necessary condition, sufficient condition*, and *necessary and sufficient condition*.

Most school textbooks explain necessary condition and sufficient condition as follows:

When we have two conditions $p(x)$ and $q(x)$, and for all x
$$p(x) \Rightarrow q(x) \text{ holds,}$$

we call $q(x)$ a necessary condition for $p(x)$ and $p(x)$ a sufficient condition for $q(x)$.

When both $p(x) \Rightarrow q(x)$ and $q(x) \Rightarrow p(x)$ hold or, in other words, when $p(x) \Leftrightarrow q(x)$ holds, we call $p(x)$ a necessary and sufficient condition for $q(x)$. In this case, we also say that the two conditions $p(x)$ and $q(x)$ are equivalent.

After giving such explanations, the textbooks provide exercises requiring students to judge which case is a necessary condition, a sufficient condition, or a necessary and sufficient condition for pairs of concretely given $p(x)$ and $q(x)$.

Accordingly, students experience only the judging of the truth of given propositions; they are not involved in active situations where they have to figure out by themselves what is the antecedent or consequent of a proposition. Perhaps because of this practice, many teachers point out that despite repeated lessons on the definition of necessary condition and sufficient condition, students do not fully understand them. Full understanding may be obtained only through experiences in which students think through various cases and try to determine whether a proposition holds true for certain antecedents and consequents.

As a prerequisite to the topic of necessary condition and sufficient condition, this lesson was designed to help students learn about the diversity of possible antecedents and consequents for hypothetical propositions.

Expected Responses and Discussion of Them

Examples of expected responses

Problems 1 and 2 are used as a warm-up exercise for problem 3. Examples of expected responses are shown here only for the propositions in problem 3.

A. If one of x or y is odd and the other is even, then _____ is odd.

$x \pm y$

$(x \pm y)^n$, where n is a natural number not less than 2

$a(x \pm y)$, where a is odd

$x \pm y + k$, where k is even

$x^m \pm y^n + k$, where k is even

$ax \pm by$, where a and b are odd

$ax^n \pm by$, where a and b are odd

$xy + k$, where k is odd

B. If _____ is odd, then one of x or y is odd and the other even.

$x \pm y$

$(x + y)^n$, where n is a natural number not less than 2

$a(x \pm y)$, where a is odd

$x \pm y + k$, where k is even

$x^m \pm y^n + k$, where k is even

$ax \pm by$, where a and b are odd

$ax^n \pm by$, where a and b are odd

C. If both x and y are odd, then _____ is odd.

xy

$x^m y^n$, where m and n are natural numbers

axy, where a is odd

$xy + k$, where k is even

$x \pm y + k$, where k is odd

$ax \pm by$, where a, b are odd and even, respectively

$ax^m \pm by^n$, where a, b are odd and even, respectively

$(x \pm y)^n + k$, where k is odd

$x^m \pm y^n + k$, where k is odd

D. If _____ is odd, then both x and y are odd.

xy

$x^m y^n$, where m and n are natural numbers

axy, where a is odd

$xy + k$, where k is even

Discussion of the responses

Problem 1 is a warm-up problem in which students judge whether a given proposition is true or not. Its purpose is to help students understand that propositions may be true or false according to differences in their antecedents or in their consequents, by proposing propositions that have the same antecedent but different consequents (propositions A, B, and C) and, conversely, those that have the same consequent but different antecedents (D, E, F). As for propositions G–J, in which the antecedents and consequents are interchanged, the purpose is for students to reconfirm what they have learned about the converse of a proposition.

In problem 2, students are led first to determine whether the sum, difference, or product of x and y fits as a consequent of a true proposition, and next to consider expressions of various combinations of x and y. It may serve as an introduction to problem 3.

In problem 3, students are led to consider suitable diverse expressions as an antecedent or a consequent. The fundamental expressions are the sum, the difference, and the product of x and y, and all their combinations are possible candidates. Therefore, although the expected responses are classified into types, strictly speaking there would not necessarily be a qualitative difference among types. When expressions proposed by the students have different forms, they should be regarded as different. It seems better to refrain from a hasty generalization or categorization.

What is considered as a further development

Judging whether a proposed expression is suitable as an antecedent or a consequent is not as easy as expected. This is true for expressions as an antecedent such as in propositions B or D in problem 3.

Substituting numerical values for variables may overcome these difficulties. Such difficulties also provide an opportunity to introduce the reduction to absurdity or the method of conversion. If students have already learned these methods of reasoning, problem 3 presents an opportunity to deepen their understanding by applying the methods.

Figure 6.35 summarizes the cases in problem 3 according to whether the proposition holds or not.

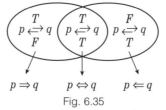

Fig. 6.35

On the basis of this lesson, the teacher can develop the definitions of necessary condition, sufficient condition, necessary and sufficient condition, and equivalence in the next lesson.

Record of the Classroom Teaching

Teaching the lesson

The lesson was taught during one period before the topic "necessary condition and sufficient condition."

1. After distributing the worksheets, the teacher had students think about problem 1. (5 minutes)

2. The teacher asked students to present their own ideas and to confirm the correct answer. Then, the teacher made sure that students understood the following points:

   - Even when statements have the same antecedents, they may be true or false according to the difference in their consequents.

   - Even when statements have the same consequents, they may be true or false according to the difference in their antecedents.

   - Propositions G and H (and also propositions I and J) are mutually converse, and the converse of a proposition is not necessarily true if the original proposition is true. In addition to the converse, it is possible to formulate the inverse and contrapositive of a proposition. (5 minutes)

3. The teacher explained problem 2. The students considered various expressions by addition, subtraction, multiplication, and their combinations. (5 minutes)

4. The students presented the expressions that they had considered, and the teacher helped the class decide whether the expressions were true or false. (7 minutes)

5. The students considered problem 3 and submitted their worksheets after they had written down their responses. (15 minutes)

6. The students presented the expressions they had produced, and then the teacher discussed with all students whether the expressions were true or had a counterexample. (13 minutes)

Student responses

The following expressions were produced by students as the lesson progressed.

Problem 2

$x + y$, $x - y$, xy, $(x + y)^2$, $(x - y)^2$, $(xy)^2$, $2xy$, $x + y + 2$, $|x - y|$, $x^2 - y$, $x^2 + y^2$, $x^2 - y^2$, $(\sqrt{xy}\,)^2$, $x(x + y)$, $xy(x + y)$, $3x + y$, $2x^2 + y$, $5x(x + y)^2 + xy$

Problem 3

A	B	C	D
$x + y$	$3(x + y)$	xy	xy
$x - y$	$x + y^3$	$(xy)^2$	x^2y^2
$(x + y)^2$	$(x + y)(x - y)$	$x - y + 1$	x^2y
$3(x + y)$	$3x + y$	$2x + y$	$3xy$
$x^3 - y^3$	$(x + y)/2$ (false)	$x^2 + y^2 - 1$	$x(y + 2)$
$xy + 1$		$x^2 + 3x + y$	$2x + y$ (false)
		$|x^2 - y^2|$ (false)	

Table 6.3 summarizes, by type, the responses for problem 3 on forty-four students' worksheets.

Proposition	Type of Response	Number of Responses
A	$x \pm y$	44
	$(x \pm y)^n$, where $n \geq 2$	29
	$a(x \pm y)$, where a is odd	14
	$x \pm y \pm k$, where k is even	7
	$x^m \pm y^n + k$, where k is even	30
	$ax \pm by$, where a, b are odd	12
	$ax^m \pm by$, where a, b are odd	13
	$xy + k$, where k is odd	11
	Others	6
B	$x \pm y$	43
	$(x \pm y)^n$, where $n \geq 2$	17
	$a(x \pm y)$, where a is odd	11
	$x \pm y + k$, where k is even	7
	$x^m \pm y^n + k$, where k is even	22
	$ax \pm by$, where a, b are odd	8
	$ax^m \pm by$, where a, b are odd	12
	Others	3
C	xy	41
	$x^m y^n$, where m, n are natural numbers	29
	axy, where a is odd	12
	$xy + k$, where k is even	3
	$x \pm y + k$, where k is odd	12
	$ax \pm by$, where one of a, b is odd, the other even	18
	$ax^n + by$, where one of a, b is odd, the other even	4
	$(x \pm y)^n + k$, where k is odd	8

Proposition	Type of Response	Number of Responses
	$x^m \pm y^n + k$, where k is odd	3
	Others	6
D	xy	39
	$x^m y^n$, where m, n are natural numbers	23
	axy, where a is odd	9
	$xy + k$, where k is even	4
	Others	2

Table 6.3
Number of each Type of Response for Problem 3

Remarks after the lesson

High-achieving students made few responses. The reason may be that even when the students thought of many candidate expressions for an antecedent or consequent, they were puzzled as to whether the answers were suitable, perhaps because they could find no essential differences among the answers.

Furthermore, in problem 3, there were fewer responses and more wrong answers for propositions B and D than for propositions A and C. As expected, creating expressions to fit a consequent is more difficult than creating expressions to fit an antecedent. It is natural that fewer response types exist for given antecedents, so judging their truth is not easy. One approach to help students overcome these difficulties may be to teach the reduction to absurdity or the method of conversion. This lesson is very effective in motivating the students to learn such topics.

Students found great pleasure in these activities, which involve individually thinking of various expressions using their own free or natural ideas, presenting them to all the students, discussing them with the other students, and formulating the new concepts called necessary condition and sufficient condition from the discourse. The open-ended teaching approach is an effective mathematization process that can be used at the introductory stage of teaching concepts. By using this approach, students realize a way to mathematize their freely proposed ideas, classify them, and gradually develop the ideas into mathematical principles or laws.

A Similar Problem

Integral Expressions

When a, b, c, and d are consecutive integers in that order, $bc - ad = 2$ always holds. Following this example, write as many integral expressions in a, b, c, and d as possible so that 0, 1, 2, or 3 appear to the right of the equals sign, as follows:

1. When the right side of the expression is 0
2. When the right side of the expression is 1
3. When the right side of the expression is 2
4. When the right side of the expression is 3

EQUATIONS OF STRAIGHT LINES

SHŪEI SASAKI
Yamagata Minami Upper Secondary School, Yamagata

The Problem and Its Context

The problem

1. Draw the graph l of an equation $2x + y - 1 = 0$, in red.
2. Draw the graphs of the following equations in black, on the coordinate system used in step 1.

   a. $2x - y + 5 = 0$ b. $x - 2y + 7 = 0$ c. $x - 2y - 3 = 0$
   d. $x + y - 5 = 0$ e. $x + y + 1 = 0$ f. $2x + y - 5 = 0$
   g. $2x + y + 5 = 0$

3. Examine the graphs and list various properties of the line l.
4. Among the lines a–g, some have a property in common with l. List the properties that line l has in common with at least two of the lines a–g, indicate the lines by their letters.
5. List the properties of the lines for a–g that are common among at least two of them. Indicate the lines by their letters.

Pedagogical context

This problem is used as an introduction to the topic "equation of a straight line" in the first grade of upper secondary school. At first, the teacher asks the students to draw graphs of several linear equations to review learning in the lower secondary school. Next the teacher has students list various common properties among the graphs. The students organize the proposed properties and extract the important ideas used to determinine a straight line. Finally the teacher clarifies how to derive equations of straight lines from these ideas.

Figure 6.36 shows the lines that students are asked to draw in steps 1 and 2 of the problem.

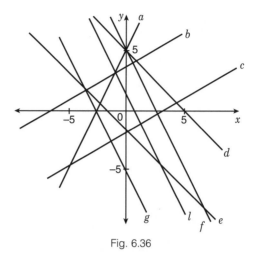

Fig. 6.36

Since accurate graphs are needed in steps 3–5, it is important that the teacher confirm the accuracy of the students' drawings. In teaching graphs of linear equations, the lesson usually ends by drawing graphs. But in this lesson, drawing graphs is the starting point of learning; this is an important feature of this approach.

In step 3, the students list properties of line *l*. In step 4, students must determine whether other lines have a property in common with *l*, and then arrange and verify their viewpoints. In step 5, students find other properties that are common among lines a–g—but not with line *l*—and once again classify and verify their viewpoints. Thus, students are led to deduce how to determine a straight line.

Expected Responses and Discussion of Them

Examples of expected responses

Because the problem can be approached intuitively by looking at graphs, the students will propose a variety of properties. The common properties can be organized according to the following five viewpoints:

Type of Response	Example
1. Slope	The slopes are negative: *d, e, f, g*
2. Point	The lines pass through the point $(-1, 3)$: *a, b*
3. *x*-intercept	The *x*-intercepts are positive: *b, d, f*
4. *y*-intercept	The *y*-intercepts are positive: *a, b, d, f*
5. Quadrant	They do not pass through the third quadrant: *d, f*

Discussion of the responses

In step 3, students may be puzzled at first when they are asked to list the properties of the line *l*. In this case, the teacher must guide the students' thinking. The teacher must have the students examine whether there are any properties among those proposed that are equivalent, though their expression may be different. For example, "the line being downward to the right" and "its slope being negative" are the same.

In step 4, students determine whether graphs exist that have the same property as each of the properties of the line l considered in step 3. For example, the slope of the line l is -2; the lines having the same slope are f and g. In this manner, several common properties will emerge, which should be classified into the five categories listed in the section "Examples of expected responses."

In step 5, students study the graphs from the five viewpoints they learned in step 4. Students are led to find properties that are different from those of line l, such as "all of the y-intercepts of a, d, and e are 5."

It should be emphasized that during this process, the teacher must give careful and serious consideration to responses by low-achieving students whose responses in ordinary lessons are not noticed or acknowledged. Through open-ended lessons like this one, the teacher should try to give these students experience in expressing their ideas and finding pleasure in doing so.

Possibility of further development

Before going further, the following points should be confirmed.

- We have *slope*, *point*, *x-intercept*, and *y-intercept* as important ideas to determine straight lines.
- In the problem posed, no straight line exists that is parallel to the y-axis.

The following equations of straight lines can be derived from what is given:

Given slope and y-intercept	$\Rightarrow y = ax + b$
Given slope and a point	$\Rightarrow y - y_0 = a(x - x_0)$
Given two points	$\Rightarrow y - y_1 = (y_2 - y_1)(x - x_1)/(x_2 - x_1)$
Given x- and y-intercepts	$\Rightarrow x/a + y/b = 1$
Parallel to the y-axis	$\Rightarrow y = c$

Further development of the problem begins with the question "Can we now find common features of equations as we did for graphs?" The teacher shows students the relation between the linear equation and its graph; i.e., equation $\Leftrightarrow$ graph. The teacher now focuses attention on finding common properties or features from the form of the equations. Then the teacher has the students examine the relations between two straight lines, such as the lines being parallel, perpendicular, or symmetric.

An open-ended lesson should provide a natural link to the lesson that follows. In this lesson, at first students find various properties and present them to the whole class. Their presentations are acknowledged individually, discussed, examined, summarized by all students, and then linked to a new topic in mathematics. That is, new learning of mathematics develops on the basis of the students' own ideas. In this way, students become aware of their participation in learning, which is a positive characteristic of this teaching approach.

RULES IN A TABLE OF A GROUP

ZENKŌ OZAWA
Yamakita Upper Secondary School, Kanagawa

The Problem and Its Context

The problem

The transformations that superimpose a regular triangle on itself (fig. 6.37) are as follows:

e: the identical transformation (no change)

w: the rotation of 120 degrees about its center O (counterclockwise)

w^2: the rotation of 240 degrees about its center O

p: the symmetric transformation (flip) with respect to the line l

q: the symmetric transformation (flip) with respect to the line m

r: the symmetric transformation (flip) with respect to the line n

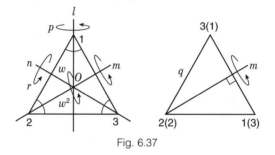

Fig. 6.37

The product of two transformations $(p \cdot q)$ of p and q is defined as the transformation that results by applying p, followed by q. For example, $p \cdot q = w^2$.

1. Complete the operation table (fig. 6.38) for $X \cdot Y$. (Fig. 6.38a shows the completed table.)

X	Y					
	e	w	w^2	p	q	r
e						
w						
w^2						
p						
q						
r						

Fig. 6.38
Table for $X \cdot Y$ as Presented to Students

X	Y					
	e	w	w^2	p	q	r
e	e	w	w^2	p	q	r
w	w	w^2	e	q	r	p
w^2	w^2	e	w	r	p	q
p	p	r	q	e	w^2	w
q	q	p	r	w	e	w^2
r	r	q	p	w^2	w	e

Fig. 6.38a
Answers for Table for $X \cdot Y$

2. Examine the operation table and list as many properties of the $X \cdot Y$ operation as possible.

Pedagogical context

This lesson's aim is to help students understand an algebraic structure. It can provide a review of the concept of a set being closed with respect to an operation or of the commutative law with respect to an operation, which students learned in the lower secondary school. It can also be used as a further development in the study of a mathematical group in the upper secondary school. For more advanced students, the lesson can serve as an introduction to the concept of a subgroup. (*Translator's note:* Since the 1980s, most of the topics mentioned here have disappeared from the curriculum.)

Although this lesson can be used to introduce the concept of a group, teaching it after teaching the concept of a group may elicit more student responses and help to deepen students' understanding.

The lesson is taught in two sessions. In the first session, students complete the operatiuon table for $X \cdot Y$. In the second session, they identify and classify properties of the operation table (and thus of the operation $X \cdot Y$).

Expected Responses and Discussion of Them

Examples of expected responses

   Typical responses

1. The sets $\{e\}$, $\{e, w, w^2\}$, $\{e, p\}$, $\{e, q\}$, $\{e, r\}$ and $\{e, w, w^2, p, q, r\}$ are all closed with respect to the operation.

2. In every set in response 1, there is a unit (identity) element $\square$ such that $\square \cdot x = x \cdot \square = x$ for all x.

3. In every set in response 1, for each element x there is an (inverse) element $\square$ such that $\square \cdot x = e$ and $x \cdot \square = e$.

4. In every set in response 1, $(x \cdot y) \cdot z = x \cdot (y \cdot z)$ holds (the associative law).

5. Every set in response 1 is a group.

6. Let $H = \{e, w, w^2\}$. Then, from $p \cdot e = p, p \cdot w = r, p \cdot w^2 = q$, we have $G = \{e, w, w^2, p, q, r\} = \{e, w, w^2\} \cup \{p, q, r\} = H \cup pH$, where $pH = \{p \cdot e, p \cdot w, p \cdot w^2\} = \{p, q, r\}$.

Furthermore, $G = H \cup qH = H \cup rH$.

7. Let $K = \{p, q, r\}$. Then $pK = Kp = qK = Kq = rK = Kr = H$; similarly, $pH = Hp = qH$ $= Hq = rH = Hr = K$, and $G = eG = Ge = wG = Gw = w^2G = Gw^2 = pG = Gp = qG = Gq$ $= rG = Gr$. (*Editor's note: $pK = Kp$ does not imply the property of commutativity; rather, it states that the resulting sets are identical.*)

8. In the operation table, each transformation appears the same number of times.

9. $e \bullet e = p \bullet p = q \bullet q = r \bullet r = e$, whereas $w \bullet w \neq e$, $w^2 \bullet w^2 \neq e$.

10. There are groups in which the commutative law holds, such as $\{e, w, w^2\}$, $\{e, p\}$ and others; and those in which the commutative law does not hold, such as $\{e, w, w^2, p, q, r\}$.

Patterns in the operation table

11. The arrangements of elements in the first row and the first column are the same.

12. The arrangements of elements in the first row and the sixth row are opposite to each other. This property also holds for the second and fourth rows, and for the third and fifth rows.

13. The lines joining the same elements in the upper and lower parts of the operation table divided by the line l (fig. 6.39) are symmetric with respect to the line l.

14. The same elements are arranged according to the directions shown by the arrows in figure 6.40.

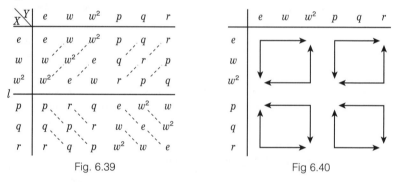

Fig. 6.39　　　　　　　　　　　　　Fig 6.40

15. If e is represented by $\bigcirc$, and p by $\triangle$, we obtain figure 6.41. A similar pattern is found with respect to w and q, and w^2 and r.

16. When the transformations in the first row are moved as shown in figure 6.42a, we obtain the third row. Moving again in the same way, we obtain the second row. A similar observation can be made for columns. (fig. 6.42b)

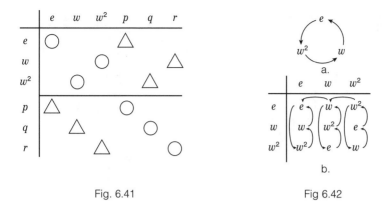

Fig. 6.41   Fig 6.42

17. In figure 6.43, A and B, C and D, E and F, and H and G are symmetric with respect to the oblique lines in the figure.

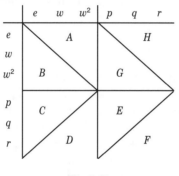

Fig. 6.43

Mixtures of patterns in the operation table

18. All transformations appear in each row and each column once and only once.

19. The entries in the operation table for the sets $\{e\}$, $\{e, w, w^2\}$, $\{e, p\}$, $\{e, q\}$, and $\{e, r\}$ are arranged symmetrically with respect to the main diagonal of the table (the commutative law).

20. In figure 6.44, a and d are entries on the main diagonal and vertices of a square of size 2; b and c are entries on the other vertices of the square. Then we have $a \cdot d = b \cdot c$.

21. In figure 6.45, a and d are entries on the main diagonal and vertices of a square of size 3; b and c are entries on the other vertices of the square. Again we have $a \cdot d = b \cdot c$.

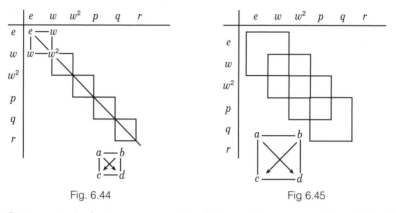

Fig. 6.44   Fig 6.45

22. Statements similar to responses 20 and 21 hold for squares of size 4, 5, or 6.

23. The commutative law does not hold in the sets $\{p, q\}$, $\{p, q, r\}$, $\{q, r\}$, and $\{e, w, w^2, p, q, r\}$. The entries are not symmetrically arranged with respect to the main diagonal.

Discussion of the responses

Even elementary school students can find some patterns in the arrangement of entries in the operation table. In the lower secondary school, this problem can help students find that an operation is closed in a subset or confirm the existence of the unit element and the inverse element in connection with learning the structure of number systems.

When the problem is used at the upper secondary school level, the lesson can focus on the idea of the subgroup, and having students compare the problem in this lesson with operations on matrices. One approach is to have students make the tables (figs. 6.46 and 6.47) for the operations $X \cdot Y$ and $X + Y$ on the following matrices. The students can then identify properties by comparing these tables with the operation table in the original problem (fig. 6.38).

$$I = \begin{pmatrix} 1 & 0 \\ 0 & 1 \end{pmatrix} \qquad A = \begin{pmatrix} 1 & 0 \\ 0 & -1 \end{pmatrix} \qquad B = \begin{pmatrix} -1 & 0 \\ 0 & 1 \end{pmatrix} \qquad C = \begin{pmatrix} -1 & 0 \\ 0 & -1 \end{pmatrix}$$

X	Y			
	I	A	B	C
I	I	A	B	C
A	A	I	C	B
B	B	C	I	A
C	C	B	A	I

Fig. 6.46
Table of $X \cdot Y$

X	Y			
	I	A	B	C
I	$2I$	Q	R	0
A	Q	$2A$	0	T
B	R	0	$2B$	H
C	0	T	H	$2C$

Fig 6.47
Table of $X + Y$

Evaluation of the problem and further developments

1. It may also be interesting for the students to consider what will be found for a rec-

tangle, a square, or other shapes. The present discussion is based on a regular triangle. (See fig. 6.37).

2. Similarly, students may be led to understand that the permutation group

$$\left\{ \begin{pmatrix} 1 & 2 & 3 \\ 1 & 2 & 3 \end{pmatrix}, \begin{pmatrix} 1 & 2 & 3 \\ 2 & 1 & 3 \end{pmatrix}, \begin{pmatrix} 1 & 2 & 3 \\ 3 & 2 & 1 \end{pmatrix}, \begin{pmatrix} 1 & 2 & 3 \\ 1 & 3 & 2 \end{pmatrix}, \begin{pmatrix} 1 & 2 & 3 \\ 2 & 3 & 1 \end{pmatrix}, \begin{pmatrix} 1 & 2 & 3 \\ 3 & 1 & 2 \end{pmatrix} \right\}$$

is isomorphic to G, or that the multiplication with A, B, C, I (fig. 6.46) is isomorphic to the composition of mappings in a coordinate plane that consist of the following four elements:

i: the identity

f: symmetry with respect to the x-axis

g: symmetry with respect to the y-axis

h: symmetry with respect to the origin

3. For advanced students, the concept of a residue class group may be introduced as another development.

4. Further development can also be based on the fact that a subgroup may be commutative or noncommutative.

For reference, the group for this problem is nothing but the dihedral group D_3, which is noncommutative and is produced by two generators a and b having the properties $a^3 = b^2 = (ab)^2 = e$.

A Similar Problem

Problem of One-Way Traffic

As shown in figure 6.48, two roads on the banks of a river have one-way traffic in opposite directions. Many bridges with two-way traffic are built at equal distances. Cars are allowed to travel in two ways: (a) they can travel on a road m units from one bridge to another, where m is a nonnegative integer; and (b) they can cross a bridge.

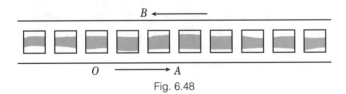

Fig. 6.48

Let a point O on one of the roads be the origin; A and B represent other points on the roads; and $\overrightarrow{OA}$ mean the movement from O to A. By $\overrightarrow{OC} = \overrightarrow{OA} + \overrightarrow{OB}$ we mean the movement from O to A, followed by the movement from A to C in the same way as that from O to B. What structure does this addition have?

A BEAUTY CONTEST FOR PETS

SHŪEI SASAKI
Yamagata Minami Upper Secondary School, Yamagata

The Problem and Its Context

The problem

The judges in a beauty contest for pets gave the following results: first place, Aoki; second place, Ito; third place, Uno; fourth place, Eguchi; fifth place, Ogata.

Before the judges' announcement, the audience was asked to guess the order of the contestants and were promised prizes according to how near their guesses were to the judges' results. The table shows four guesses by A, B, C, and D together with the judges' decision.

Contestant	Aoki	Ito	Unmo	Eguchi	Ogata
Final result	1	2	3	4	5
Guess by A	3	1	4	5	2
Guess by B	2	3	1	4	5
Guess by C	5	3	2	4	1
Guess by D	4	2	3	1	5

Fig. 6.49

How shall we decide to rank the guesses of A, B, C, and D? Write as many different methods of ranking as you can. For each method, determine who among A, B, C, and D will rank first, second, and third. Write their letters in the order in which they are ranked.

Pedagogical context

The problem requires students to consider ways of ranking guesses. In other words, it requires them to devise methods to measure the degree of nearness between the guesses and the final result. Thus, it may be called a problem of how to measure. Its purpose is to help students understand the relation between assumption and conclusion and appreciate the importance of flexible thinking. It clarifies the following two facts:

- The possible scales or methods of measurement are numerous and diverse.
- The ranking may differ according to the scale applied.

The problem may be used at any grade level or at any time during the school year as a special topic. Though responses may differ according to students' grade levels or what they have learned, such differences themselves are interesting.

The following approach could be used to teach the lesson:

The teacher prepares pictures of five beauty contestants and displays them on the blackboard. Students rank them freely according to their taste and present their opinions. Next the teacher announces the judges' results and says to the students, "We would like to compare your rankings of the contestants according to their nearness to the judges' results.

How can we measure the nearness? There seems no other way than expressing the nearness by a number. So let us consider methods or rules by which to express the degree of nearness as a number."

The teacher may first have students tackle the problem. A little later, it may become necessary to explain how to respond. The teacher then asks some students to present their rules and the rankings they computed using their rules. Then the teacher encourages other students to follow this example in their presentations.

Expected Responses and Discussion of Them

Examples of expected responses

Many rules will be proposed and may be classified as follows:

1. The total number of coincidences between the judges' and students' rankings, in decending order

2. The sums of weights given to each coincidence between the judges' and the students' rankings of first, second, third, ...; in descending order

3. The sums of absolute differences (where *absolute difference* means $|a - b|$) or squares of differences between the judges' and student rankings; in ascending order

4. The sums of points obtained by subtracting the absolute differences between the judges' and students' rankings from a specified number; in descending order (a demerit system)

5. A modification of methods 3 or 4 that ignores differences that are larger than a specified number

6. The numbers of times that two student ranks (e.g., second and third) must be interchanged until the student ranking matches the judges' ranking; in ascending order

7. The sums of the student ranks of the contestants who were ranked first, second, or third by the judges; in ascending order (or the sums of student ranks of contestants who were ranked fourth or fifth by the judges; in descending order)

8. The ranks given to a particular contestant; in ascending order

9. The judges' ranks of the contestants who were first in the students' rankings; in ascending order

10. The sum of points defined in such a way that points 1, 2, 3, and 4 are given in each student's ranking according to the ascending order among students of absolute difference between judges' and students' ranking; in ascending order

11. A modification of method 3 that ignores the largest and smallest absolute differences between the judges' and students' rankings (like the scoring in a gymnastics competition)

12. The student rankings of contestants arranged in order of the judges' rankings (e.g., for student A, the numbers would be 31452); in descending order

13. The numbers of coincidences between the judges' and students' rankings for each pair of contestants (e.g., if the judges and the student both ranked Ito before Uno, there is a coincidence); in descending order

14. Others

Discussion of the responses

After students clearly understand the meaning of the problem and what is expected of them, it is important for the teacher to let them consider the problem freely by themselves. Next, during their presentations, the teacher should help students see how diverse their methods of measuring are. A student may notice a new method from a suggestion in another student's presentation. Thus, the teacher should try to have as many presentations as possible. After modifying students' poor expressions of their ideas, if necessary, and grouping those based on the same principle or feature, the teacher should summarize and classify the presentations into the types listed in the examples of expected responses.

The computation of the rankings according to each rule found by students should follow the summary. For example:

1. If students are ranked using method 1, we have D, B, C because A:0, B:2, C:1, D:3

2. If method 3 is used, we have B, D, A:

$$A: \quad 2 + 1 + 1 + 1 + 3 = 8$$
$$B: \quad 1 + 1 + 2 + 0 + 0 = 4$$
$$C: \quad 4 + 1 + 1 + 0 + 4 = 10$$
$$D: \quad 3 + 0 + 0 + 3 + 0 = 6$$

3. Method 12 yields B, A, D because A:31452, B:23145, C:53241, D:42315.

Thus, students may understand that if the scales used to measure nearness differ, then the rankings according to the scales may differ.

Next comes the evaluation of the rules. Which among the variety of student-devised rules are better than others? The primary aim of the problem is to have students think of as many rules for measurement as possible, but its secondary aim is to have students think of qualitatively superior rules.

Qualitatively superior rules may mean those that satisfy the following two criteria:

- The rule relates to all five contestants, but not to any specific one.
- The degree of difference in each case is measured numerically.

According to these criteria, methods 3, 4, 6, 10, and 13 in the example of expected responses are qualitatively superior. Methods 1, 7, 8, and 9 are qualitatively inferior.

Further development

Although the lesson will usually end at the stage above it is also possible to develop Spearman's rank-order correlation from method 3 of the sum of squares or Kendall's rank-order correlation from method 13.

A Similar Problem

Measuring the Degree of Curvature in a Road

Figure 6.50 shows curved roads having a circular arc from *A* to *B*, in the same scale.

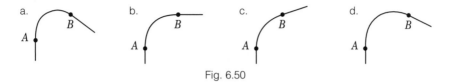

Fig. 6.50

Both sharp and more gradual curves exist. For example, the curve in *a* seems to be sharper than the curve in *d*. We would like to express the sharpness of the curves from *A* to *B* by a number. How shall we do it? There are many methods to express it numerically. Write as many of them as possible. If you find it difficult to express your methods in words, you may explain by drawing figures.

Chapter 7

Round-Table Discussion to Review the Project

DATE: February 15, 1977

PLACE: National Institute for Educational Research (Tokyo, Japan)

ATTENDANTS: Shigeru Shimada (Presider)   National Institute for Educational Research
                Toshio Sawada (Recorder)   National Institute for Educational Research
                Yoshishige Sugiyama   Tokyo Gakugei University
                Yoshihiko Hashimoto   National Institute for Educational Research
                Nobuhiko Nohda   Fukui University
                Yoshio Takeuchi   Yamagata University
                Hiroshi Kimura   Yokohama National University
                Kanjiro Kobayashi   The Elementary School at Chiba University
                Shūei Sasaki   Yamagata Minami Upper Secondary School
                Yukio Yoshikawa   The Lower and Upper Secondary School at
                                                the University of Tokyo

SHIMADA: We have heard opinions and impressions from teachers who have taught the open-ended problem-solving lessons in their classrooms. They pointed out several merits of this teaching style. For example, it was very enjoyable; students were able to find a variety of ideas in accordance with their abilities by determining their own viewpoints for the situations; as students expressed unexpected ideas, the teacher discovered new aspects of those students' ways of thinking that had previously not been apparent; in upper secondary schools, students of average or lower ability, who usually were not eager to express their opinions, now did so and actively participated in the lessons.

On the other hand, some demerits were also reported. For instance, at times it was difficult for the teachers to devise appropriate wording necessary to pose the problems or to ask questions of the students; it was difficult to summarize many ideas of the students in a suitable way at the end of the lessons, and to further develop those ideas afterwards.

To sum up, teachers and students can enjoy the open-ended approach in the classroom; however, the teacher needs the ability to teach using this approach. Now I would like you to talk about some possible future problems, such as how to incorporate this teaching style into the whole plan of teaching school mathematics, how to evaluate it, and how to develop it further.

HOW TO EVALUATE

SHIMADA: How did you check the merits of this approach? How did you evaluate reactions of the students who learned through the open-ended approach? Mr. Hashimoto.

HASHIMOTO: I think that there are two objects of the evaluation. One is the evaluation of students' learning, and the other is the evaluation of classroom teaching itself. For each, it is important to evaluate both quantity and quality.

First, I will explain how we evaluated students' learning in the classroom. We posed one open-ended problem to the students at the beginning of the classroom lesson. When the students worked on the problem individually, we asked them to write their solutions on their worksheets. Then we gathered these papers. After the lesson, we checked the students' written responses. Our viewpoints were as follows: How many solutions did the students find? How many different ideas did they come up with? In the sense of mathematical quality, how rich were their ideas—that is, were they superficial or deeper than usual in a mathematical sense? Sometimes we returned the worksheets to the students at the beginning of the next lesson and had them discuss their own ideas and those of others.

Next, I will explain the other aspect of the evaluation. We conducted the open-ended approach to teaching three times during about three months, only in the experimental classrooms. At the beginning and the end of this three-month period, we gave open-ended problems to the students to solve. We quantitatively and qualitatively examined the changes in students' responses through the differences between their scores on the pretests and the posttests. We also compared these results with those of students who were not taught using this approach. We used three measures of students' responses, which were described in chapter 3: the total number of responses, or fluency; the total number of positive responses, or flexibility; and the total number of weighted positive responses, or originality. The first two measures were regarded as quantitative ones, whereas the last was qualitative. We studied the students' growth using lessons based on this two-dimensional scale, and found that those who learned using open-ended problems received higher scores on the tests than students who did not, from the perspectives of both quantity and quality.

As another evaluation, we gave a number pair (P,Q) to each student's set of responses, where P is the number of positive responses, and Q is the maximum weight of the positive responses. We compared the preteaching and postteaching number pairs. Then the changes in these number pairs for each student were classified into two groups, one being favorable, the other unfavorable. By comparing the ratio of students in a class who showed a favorable change, we obtained results similar to those of the other evaluations.

SHIMADA: Didn't you introduce somewhat different viewpoints, Mr. Nohda?

NOHDA: Yes, we did. We considered that the number of weighted positive responses was inadequate to evaluate the generality and abstractness of mathematical thinking. We classified the number of positive responses into three groups—lower, middle, and upper—from two perspectives. One perspective was the depth of the mathematical viewpoint, and the other was the generality and abstractness of a student's proposition or response. The results for each student were summarized in the form of a 3×3 matrix. (See fig. 7.1.)

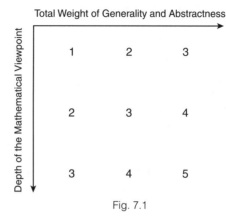

Fig. 7.1

Because the sum of the elements of this matrix is a number of positive responses, the progress of each student could be easily grasped by the transition of this matrix concerning the same problem when we gather data during and after the lesson. Furthermore, we may also get a score from a matrix by giving a weight to each element, as shown in the table. Evaluations in our group were carried out using this method. The results were similar to those of Mr. Hashimoto. However, our method would make it easier to grasp and interpret the students' transitions.

SHIMADA: Determining how to observe and evaluate students' responses is our own open-ended problem! How shall we organize the viewpoints from which we evaluate? This is the problem. "Number of positive responses," "total number of responses," and "weight for responses" are means to answer such a question. But there may be another measure, such as the degree of elegance in the students' expression, which so far we have been unable to handle, though we wanted to. In expressing mathematics, there may be excellent as well as poor expressions. The students' abilities would also be reflected in such expressions.

We need this kind of multidimensional perspective for our classrooms. Up to now, we have employed such intricate methods as the ones described today, regardless of the accompanying hardship, for the sake of scientific study. However, for practical use in the schools, we need a simplified method such as the one proposed by Mr. Nohda. I think a simplified method can be devised that will show results similar to those obtained using an intricate method.

KOBAYASHI: I have some comments on this point. When I see elementary school children's responses in the classroom, I usually find many that can be regarded as the same from a certain point of view. If I ask children to write down as many findings as possible, they are likely to redundantly write down several sentences that are based on the same ideas but expressed differently. It seems important for the teacher to devise a way to challenge children to change their viewpoints.

I think that an important suggestion to the children is "Put what you have done in order" or "Look over your list of ideas again." It is important to help children realize that their viewpoint is only one among many possibilities.

In the classroom, we should always evaluate children's responses in order to "diagnose and treat" them. We should have simple approaches for evaluating quality and quantity. This is a challenge for us.

YOSHIKAWA: I would like to give an example of evaluation that I used in my classrooms. After teaching open-ended problems for seven or eight classroom periods, I told the students that they had better neatly keep their studies on the theme that was most interesting to them, because they would be asked to write about it on a test the next week. One week later, they had to do so within a given time on the test, and I studied their responses.

HASHIMOTO: What grade were they in?

YOSHIKAWA: The eleventh grade. I tried, as a real teacher in service, to determine how we could use students' reports and responses for evaluation. I considered the problem by myself and also asked my friends' opinions. But I could not find any good way. Finally, I studied the students' responses by reading and grading their papers, and doing the same thing again a few days later. Though my criteria for grading were a bit unstable, I found that some students responded very well and offered many viewpoints in the classroom, but that they did not write well on the test because of their poor expression or lack of precision. On the other hand, some students who did not show remarkable responses in the classroom made well-organized reports.

At the end, I put all the students' reports in order and made a booklet, adding my own comments to those of the students. I distributed a booklet to each student. Students could then see the responses and viewpoints of the other students.

SHIMADA: Though it may sound curious, one purpose of our study is to increase the opportunity to praise students for their insights and activities. As a way of evaluation, we wish our praise to be specific, such as "You had so many responses," "You thought deeply in the matter," or "You were very flexible in changing viewpoints," and so on.

HOW TO INCORPORATE THIS TEACHING STYLE INTO THE TEACHING PROGRAM

SHIMADA: Because not all everyday teaching is carried out solely by using open-ended problems, it is necessary to consider how to incorporate this teaching style into the whole teaching program. Would you share your opinions? Mr. Takeuchi.

TAKEUCHI: We have conducted this research for a few years. What I had in mind when we introduced the open-ended approach into classroom teaching is that mathematics education should aim to enhance all aspects of mathematical activities. From that viewpoint, we find that our traditional school mathematics teaching has been similar to college-type teaching, wherein instructors have been eagerly trying to teach ready-made mathematics to convey typical paradigms to be followed by students, which will likely lead to a closed view of mathematics. This teaching style has lasted since about the nineteenth century.

In contrast to that system, we have another system, which might be called a kindergarten type. This type existed before the current typical paradigms were established. Historically speaking, it originated in a very productive stage of the sixteenth to the seventeenth or eighteenth centuries. The teachers had the freedom to produce anything they liked, and there were no fixed forms or typical examples. The energy came from the spirit to challenge unknown problems. The approach was, so to say, trial and error; that is, if they failed, they started again. The history of human development was a product of such continuing efforts.

When the people were confronted with one problem, they endeavored to find an answer, and finally got an answer in some way. But that answer was only a tentative solution to the problem. They soon found a new problem that the previous solution could not address. They would confront the problem again. The series of this kind of activity contributed to the growth of modern science. When it comes to school mathematics, this productive stage of science development has more significant meaning than the college type of teaching. *Author's note:* The development of science may be divided into three stages: (1) productive stage—a new paradigm or discovery is created; (2) constructive/expansive stage—theories are structured and applications are expanded; (3) "conflict" stage—defects are found and the basic philosophy is challenged.

In the latter part of our study, our group was trying to introduce the open-ended approach into the classrooms from this viewpoint. I would like to mention some of its merits. One is that we become aware of the importance of paying special attention to the interrelationships in the teaching content.

Of course, in the traditional curricula, these interrelationships have also been considered. However, they are not apparent in textbooks that describe mathematics content piece-by-piece in a linear order by dividing them into chapters and sections. In order to succeed in this approach, we need a clear understanding of the networks of content in all areas of mathematics, woven by such interrelations as warp and weft.

Teachers who taught using the open-ended approach reported that students enjoyed learning mathematics; in particular, average or low-ability students actively participated in learning, their eyes sparkling with joy. This fact has significant value in itself. Students carried out their own mathematical activity and tried to think through the problem situation individually, and more valuably, presented their own ideas to contribute to the class. In the traditional teaching style, a scene like this could hardly be realized despite teachers' intentions to do so.

Another merit was that we provided time for discussion in the classroom teaching. Students presented their ideas and approaches, and criticized those of others. Criticizing is itself an important part of what I call scientific or mathematical activities. The open-ended approach can provide opportunities for students to give their own opinions and rationales. This is a significant merit of the open-ended approach.

One teacher reported that he was puzzled about how to make a summary at the end of his teaching. In my opinion, a summary should also be open, not closed. Even when the teacher is content with having done a good summary, students may feel something has been "left hanging." And this just makes sense, since some unsolved part of the problem could be used in the next lesson as a clue for the next learning activities. In this sense, in making a summary, the teacher should not make it closed by being unduly influenced by the word *summary.*

SHIMADA: What is your opinion, Mr. Kimura?

Kimura: In the Kanagawa group, we are discussing whether an open-ended approach might be used to evaluate students' attitudes; specifically, we are considering how to evaluate such attitudes as the spirit of inquiry, perseverance, or concentration.

SASAKI: Though it might have nothing to do with the place of the open-ended approach, I would like to make a comment. Especially in high school mathematics teaching, there is usually little opportunity to have group discussion in the classroom. Even in the open-ended teaching, we had far fewer situations in which discussions were carried out than we had in the elementary schools. I think we should prepare more opportunities for group learning.

SHIMADA: When it comes to the teaching of logical reasoning in lower secondary schools, it is difficult to help students understand the need for proof or the axiomatic approach. If a student has a proposition that is different from ones offered by others, the student might want to insist that this proposition is also true. In such a case, all the students may acknowledge the need to prove it. Here the students are driven into a situation where there is a need for proof.

In a closed situation, just one proposition is proposed at a time. Students have no opportunity to present another proposition and compare the two. Thus, the need for logical inference is limited. On the other hand, in an open situation, there are more opportunities that spur students toward proof. We should help students get accustomed to such situations gradually from the elementary school onward. Ideally speaking, such opportunities would become more frequent at the senior high school level.

SUGIYAMA: I noticed an important point among Mr. Takeuchi's comments; that is, teaching through the use of open-ended problems and enhancing mathematical activities—in Mr. Takeuchi's words—involves fostering students' creativity. When students look at things from various perspectives, especially from unprescribed ones, it is a kind of discovery, and they are in the first stage of developing creativity. However, I think that not only finding various new facts is an aspect of creativity, but also systematizing them in good order. As Mr. Takeuchi mentioned, it is important in an open-ended approach to help students make a summary by themselves. If we do so, it is possible to neatly place the open-ended approach into the whole teaching plan.

SHIMADA: In all processes of mathematics teaching or mathematical activities, we can find two types of thinking that are currently in vogue—the divergent one and the convergent one. In traditional mathematics teaching, convergent thinking has dominated, and the thinking is supposed to converge on one point. However, the process of creativity must involve *both* divergent and convergent thinking. In some cases, the two must go hand-in-hand, and in other cases they must proceed separately. We may call the combination the "whole mathematical activity" in which the types of thinking shift in appropriate ways.

As we have proposed our new approach, some people might think that everything is now possible using this approach, but this is not our intention. What we maintain is that we should develop a good balance between divergent and convergent thinking, and this balance should make mathematics teaching more vivid. This is our intention.

SUGIYAMA: I appreciate our open-ended approach because it introduces a divergent way in mathematics teaching where convergent ways have traditionally dominated, and also because it tries to evaluate such a way of thinking.

KIMURA: This approach has another value in that we can help students understand how to learn mathematics.

SASAKI: Students are learning to change their viewpoints. If they fail using one method or approach, then they try another. This training would be effective for them even in solving entrance-examination problems, though it may sound somewhat beside the point.

FUTURE PROBLEMS

SHIMADA: We have been talking about methods of evaluation, the place of our approach in the whole program, and related reflections. Now, as this meeting is nearing its end, we would like to talk about future problems.

NOHDA: In our group, teachers individually summed up responses of their students in open-ended problems. In advance, we had a long meeting to clarify the way to sum up. The work of summing up was carried out by teachers after pretesting and posttesting. We found considerable differences among the teachers' evaluations of similar responses of students. This is undesirable. We should give the same score to the same responses. But how can we develop a reliable measuring scale? How should we select a reference point? This is an important problem.

KOBAYASHI: Students' responses seem to depend on the teacher. With a different teacher, students' responses change, even when the teacher uses the same open-ended problem. I think there might be an apt teaching pattern for use in the open-ended approach. We should develop an appropriate questioning pattern. Just trying this style of teaching and finding it interesting is not sufficient. We need an accumulation of, and reflection on, daily practices.

SHIMADA: It may be said that the key points in Mr. Kobayashi's comments are illustrated in the examples of questioning and lesson development in chapters 4–6.
          Frankly speaking, because of the various restrictions, the scope of topics we have tried up to now cannot help but be limited to those in the second term of the school year. So, one of the remaining problems is to collect more exemplary cases in which the guideline for posing problems and developing lessons is easy to use. In chapter 3, we provided some hints on how to make open-ended problems. Mr. Sugiyama, would you mention some points we should consider when we prepare open-ended problems?

SUGIYAMA: I think the open-ended problems we have studied up to now can be divided into three groups. The first is "how to find" problems. Here students are asked to find rules or relations. In such problems, it is not sufficient that students simply find some rules or relations; those rules or relations must have some mathematical significance. The second group is "how to classify" problems. Here classification itself must have mathematical meaning and contribute to the formation of mathematical concepts. We need to pay attention to this point. The teacher should choose a problem that will affect students, at least in the stage of summary. The third group is "how to measure" problems. The problems of how to measure the degree of curvature or of how to rank things are examples of this group. Most of them have much mathematical significance. I think they were highly effective among the open-ended problems we used.

We have to be especially careful in using the first and second kinds, because they sometimes turn out to be mere open-ended problems without any mathematical significance.

SHIMADA: Certainly the open-ended problems would be roughly classified into the three large groups as Mr. Sugiyama has said. Concerning the problems of "how to measure," it should also be noted that there is a risk of being so open that anything might be an answer. To avoid this, we should add a certain condition to the situation, so that the openness of the problem is restricted within a certain range.

It may be generally stated that those restrictive conditions should allow some significant mathematical aspects. On the other hand, the problem may extend beyond the students' reach when the mathematical aspects are of the level of college or specialist mathematics. So the restrictive conditions should also place the problem within students' reach.

In this sense, it is very hard to develop new open-ended problems. Developing new problems is a future research theme. When we find a good open-ended problem, however small, it would be desirable to share it with colleagues at every opportunity.

KOBAYASHI: I agree with you, Mr. Shimada, especially in the elementary schools. The situation of the problem must also be interesting to the students. When students are not interested in the problem situation, they may not feel motivated to study it, even though the problem is open and has significant mathematical value.

TAKEUCHI: In my previous remarks, I vaguely referred to "mathematical activity." I think one of our future tasks is to develop a pattern or scenario by which teachers can manage such mathematical activities, through the concrete analyses of teaching practices. To this end, it is important to realize the mathematical values involved and to reexamine the restrictions attached to the open-ended problem.

SUGIYAMA: Mr. Kobayashi mentioned the teaching pattern. I am wondering: What is the teacher's role? Should the teacher have some particular teaching style? In the open-ended style of teaching, asking the students only patterned questions such as "Are there any other opinions?" will not do. It is important for the teacher to ask specific questions in order to help students find some general rules. I think we should study what mental attitudes the teacher should have.

SHIMADA: Do you think that the first thing teachers must do is to solve open-ended problems by themselves?

SUGIYAMA: They should consider how the problem can be extended after solving it themselves. They should have experiences of extension and generalization of their own.

KOBAYASHI: In order to avoid misunderstanding, I would like to add to what I meant by "pattern." Teachers must ultimately search for the most suitable ways for their own teaching. A pattern is not merely a stereotyped way applicable to any teacher.

SHIMADA: One of my strongly held opinions is that there is no typical teaching pattern suitable for all teachers. Each teacher should have his or her own way of teaching that reflects his or her distinctive personality. If not, the teaching may seem inappropriate even from the students' eyes and will not last long. At the same time, it is necessary for the teacher to be acquainted with some essential points of teaching.

By the way, up to now we have talked about open-ended problems for teaching at the school levels. I think we also need to develop some open-ended problems for use in teacher training in the universities. What do you think about it?

SUGIYAMA: I agree with you.

SHIMADA: May I give an example of a problem for teacher education? Isn't it a kind of open-ended problem to ask preservice teachers to list the kinds of students' responses that they can expect for problems included in this book? I think such an exercise would be good for teacher education.

SUGIYAMA: Once I gave a problem of a water flask with triangular bases to my university students and found that their viewpoints were so poor that some answered only that "the shape of the surface would be an isosceles triangle," quite like schoolchildren; or still worse, only that "the volume of water is constant even when tilted."

SAWADA: I would like to change the topic. I think that the open-ended problems we have developed are restrictive. Under these restrictions it might be difficult to develop more problems. We know that it is difficult to popularize this teaching style, though we insist that the open-ended approach is highly effective. What if we delete "end" from "open-end"? I think that our philosophy of mathematical activities exists in this "open approach" to teaching.

Some time ago Mr. Sugioka of Chiba University mentioned that he wanted to enlarge the meaning of *problem* in our study in order to include problems whose answers are determined uniquely but whose solution method is not unique, and also include problems that have insufficient conditions. If we include these problems, we could more easily develop problem situations by modifying some problems in the textbooks and using them in our approach.

SASAKI: We have difficulty in making open-ended problems. But if we use this open approach in the classroom, we can provide many more situations. If we explain our intention to teachers and ask them to teach students using this open approach and to categorize their practices, then a new perspective will naturally emerge from there.

SHIMADA: Being open at some stage of mathematical activities, in the sense used by Mr. Takeuchi, may mean being undecided, or there being a certain degree of freedom to choose, and it does not necessarily exclude the case where one may find a unique solution irrespective of one's choice of solution method. If we allow non–multiple solution problems, what teaching styles can we prepare? An important research theme in the future would be to develop a broader concept of an "open situation." The cases described in this book were developed under the severe working condition of scientific study, with the intention of suggesting to teachers that actively using such open-ended situations in their teaching will yield good results.

Though other issues remain for discussion, we are running out of time so I must close this meeting. Thank you all very much.